Time Series Algorithms Recipes

Implement Machine Learning and Deep Learning Techniques with Python

Akshay R Kulkarni
Adarsha Shivananda
Anoosh Kulkarni
V Adithya Krishnan

Apress®

Time Series Algorithms Recipes: Implement Machine Learning and Deep Learning Techniques with Python

Akshay R Kulkarni
Bangalore, Karnataka, India

Adarsha Shivananda
Hosanagara, Karnataka, India

Anoosh Kulkarni
Bangalore, India

V Adithya Krishnan
Navi Mumbai, India

ISBN-13 (pbk): 978-1-4842-8977-8
https://doi.org/10.1007/978-1-4842-8978-5

ISBN-13 (electronic): 978-1-4842-8978-5

Managing Director, Apress Media LLC: Welmoed Spahr
Acquisitions Editor: Celestin Suresh John
Development Editor: Laura Berendson
Coordinating Editor: Mark Powers

Cover designed by eStudioCalamar

Cover image by Aron Visuals on Unsplash (www.unsplash.com)

Distributed to the book trade worldwide by Apress Media, LLC, 1 New York Plaza, New York, NY 10004, U.S.A. Phone 1-800-SPRINGER, fax (201) 348-4505, e-mail orders-ny@springer-sbm.com, or visit www.springeronline.com. Apress Media, LLC is a California LLC and the sole member (owner) is Springer Science + Business Media Finance Inc (SSBM Finance Inc). SSBM Finance Inc is a **Delaware** corporation.

For information on translations, please e-mail booktranslations@springernature.com; for reprint, paperback, or audio rights, please e-mail bookpermissions@springernature.com.

Apress titles may be purchased in bulk for academic, corporate, or promotional use. eBook versions and licenses are also available for most titles. For more information, reference our Print and eBook Bulk Sales web page at http://www.apress.com/bulk-sales.

Any source code or other supplementary material referenced by the author in this book is available to readers on GitHub (https://github.com/Apress). For more detailed information, please visit http://www.apress.com/source-code.

Printed on acid-free paper

To our families

Table of Contents

About the Authors..xi

About the Technical Reviewer xiii

Preface xv

Chapter 1: Getting Started with Time Series 1

  Recipe 1-1A. Reading Time Series Objects (Air Passengers)..................2

    Problem 2

    Solution 2

    How It Works 2

  Recipe 1-1B. Reading Time Series Objects (India GDP Data)..................4

    Problem 4

    Solution 4

    How It Works 4

  Recipe 1-2. Saving Time Series Objects 6

    Problem 6

    Solution 6

    How It Works 6

  Recipe 1-3A. Exploring Types of Time Series Data: Univariate..............7

    Problem 7

    Solution 7

    How It Works 7

Recipe 1-3B. Exploring Types of Time Series Data: Multivariate 9

    Problem 9

    Solution 9

    How It Works 10

Recipe 1-4A. Time Series Components: Trends 13

    Problem 13

    Solution 13

    How It Works 13

Recipe 1-4B. Time Series Components: Seasonality 15

    Problem 15

    Solution 15

    How It Works 15

Recipe 1-4C. Time Series Components: Seasonality (cont'd.) 18

    Problem 18

    Solution 18

    How It Works 19

Recipe 1-5A. Time Series Decomposition: Additive Model 21

    Problem 21

    Solution 21

    How It Works 22

Recipe 1-5B. Time Series Decomposition: Multiplicative Model 24

    Problem 24

    Solution 25

    How It Works 25

Recipe 1-6. Visualization of Seasonality 27

    Problem 27

    Solution 28

    How It Works 28

Chapter 2: Statistical Univariate Modeling 33

Recipe 2-1. Moving Average (MA) Forecast 34

Problem 34

Solution 34

How It Works 34

Recipe 2-2. Autoregressive (AR) Model..38

Problem 38

Solution 38

How It Works 38

Recipe 2-3. Autoregressive Moving Average (ARMA) Model 43

Problem 43

Solution 43

How It Works 44

Recipe 2-4. Autoregressive Integrated Moving Average (ARIMA) Model............49

Problem 49

Solution 49

How It Works 49

Recipe 2-5. Grid Search Hyperparameter Tuning for ARIMA Model 54

Problem 54

Solution 54

How It Works 54

Recipe 2-6. Seasonal Autoregressive Integrated Moving
Average (SARIMA) Model 60

Problem 60

Solution 60

How It Works 60

Recipe 2-7. Simple Exponential Smoothing (SES) Model 62

    Problem 62

    Solution 63

    How It Works 63

Recipe 2-8. Holt-Winters (HW) Model 64

    Problem 64

    Solution 65

    How It Works 65

Chapter 3: Advanced Univariate and Statistical Multivariate Modeling 67

Recipe 3-1. FBProphet Univariate Time Series Modeling 68

    Problem 68

    Solution 68

    How It Works 68

Recipe 3-2. FBProphet Modeling by Controlling the Change Points 73

    Problem 73

    Solution 73

    How It Works 74

Recipe 3-3. FBProphet Modeling by Adjusting Trends 79

    Problem 79

    Solution 79

    How It Works 79

Recipe 3-4. FBProphet Modeling with Holidays 82

    Problem 82

    Solution 82

    How It Works 82

Recipe 3-5. FBProphet Modeling with Added Regressors 84

    Problem 84

    Solution 84

    How It Works 84

Recipe 3-6. Time Series Forecasting Using Auto-ARIMA 87

    Problem 87

    Solution 87

    How It Works 87

Recipe 3-7. Multivariate Time Series Forecasting Using the VAR Model 96

    Problem 96

    Solution 96

    How It Works 96

Chapter 4: Machine Learning Regression–based Forecasting..........103

Recipe 4-1. Formulating Regression Modeling for Time Series Forecasting 104

    Problem 104

    Solution 104

    How It Works 104

Recipe 4-2. Implementing the XGBoost Model 112

    Problem 112

    Solution 112

    How It Works 112

Recipe 4-3. Implementing the LightGBM Model 114

    Problem 114

    Solution 114

    How It Works 114

Recipe 4-4. Implementing the Random Forest Model 116

    Problem 116

    Solution 116

    How It Works 116

Recipe 4-5. Selecting the Best Model 118

    Problem 118

    Solution 118

    How It Works 119

Chapter 5: Deep Learning–based Time Series Forecasting 127

Recipe 5-1. Time Series Forecasting Using LSTM 128

    Problem 128

    Solution 128

    How It Works 128

Recipe 5-2. Multivariate Time Series Forecasting Using the GRU Model 136

    Problem 136

    Solution 136

    How It Works 136

Recipe 5-3. Time Series Forecasting Using NeuralProphet 158

    Problem 158

    Solution 158

    How It Works 158

Recipe 5-4. Time Series Forecasting Using RNN 164

    Problem 164

    Solution 165

    How It Works 165

Index 169

About the Authors

Akshay R Kulkarni is an artificial intelligence (AI) and machine learning (ML) evangelist and thought leader. He has consulted several Fortune 500 and global enterprises to drive AI and data science–led strategic transformations. He is a Google developer, an author, and a regular speaker at major AI and data science conferences (including the O'Reilly Strata Data & AI Conference and Great Indian Developer Summit (GIDS)). He is a visiting faculty member at some of the top graduate institutes in India. In 2019, he was featured as one of India's "top 40 under 40" data scientists. In his spare time, Akshay enjoys reading, writing, coding, and helping aspiring data scientists. He lives in Bangalore with his family.

Adarsha Shivananda is a data science and MLOps leader. He is working on creating world-class MLOps capabilities to ensure continuous value delivery from AI. He aims to build a pool of exceptional data scientists within and outside organizations to solve problems through training programs. He always wants to stay ahead of the curve. Adarsha has worked extensively in the pharma, healthcare, CPG, retail, and marketing domains. He lives in Bangalore and loves to read and teach data science.

Anoosh Kulkarni is a senior AI consultant. He has worked with global clients across multiple domains to help them solve their business problems using machine learning, natural language processing (NLP), and deep learning. Anoosh is passionate about guiding and mentoring people in their data science journey. He leads data science/machine learning meet-ups and helps aspiring data scientists navigate their careers. He also conducts ML/AI workshops at universities and is actively involved in conducting webinars, talks, and sessions on AI and data science. He lives in Bangalore with his family.

V Adithya Krishnan is a data scientist and MLOps engineer. He has worked with various global clients across multiple domains and helped them to solve their business problems extensively using advanced ML applications. He has experience across multiple fields of AI-ML, including time series forecasting, deep learning, NLP, ML operations, image processing, and data analytics. Presently, he is working on a state-of-the-art value observability suite for models in production, which includes continuous model and data monitoring along with the business value realized. He presented a paper, "Deep Learning Based Approach for Range Estimation," written in collaboration with the DRDO, at an IEEE conference. He lives in Chennai with his family.

About the Technical Reviewer

Krishnendu Dasgupta is a co-founder of DOCONVID AI. He is a computer science and engineering graduate with a decade of experience building solutions and platforms on applied machine learning. He has worked with NTT DATA, PwC, and Thoucentric and is now working on applied AI research in medical imaging and decentralized privacy-preserving machine learning in healthcare. Krishnendu is an alumnus of the MIT Entrepreneurship and Innovation Bootcamp and devotes his free time as an applied AI and ML research volunteer for various research NGOs and universities across the world.

Preface

Before reading this book, you should have a basic knowledge of statistics, machine learning, and Python programming. If you want to learn how to build basic to advanced time series forecasting models, then this book will help by providing recipes for implementation in Python. By the end of the book, you will have practical knowledge of all the different types of modeling methods in time series.

The desire to know the unknown and to predict the future has been part of human culture for ages. This desire has driven mankind toward the discipline of forecasting. Time series forecasting predicts unknown future data points based on the data's previous (past) observed pattern. It can depend not only on the previous target points and time (univariate) but also on other independent variables (multivariate). This book is a cookbook containing various recipes to handle time series forecasting.

Data scientists starting a new time series project but don't have prior experience in this domain can easily utilize the various recipes in this book, which are domain agnostic, to kick-start and ease their development process.

This book is divided into five chapters. Chapter 1 covers recipes for reading and processing the time series data and basic Exploratory Data Analysis (EDA). The following three chapters cover various forecasting modeling techniques for univariate and multivariate datasets. Chapter 2 has recipes for multiple statistical univariate forecasting methods, with more advanced techniques continued in Chapter 3. Chapter 3 also covers statistical multivariate methods. Chapter 4 covers time series forecasting using machine learning (regression-based). Chapter 5 is on advanced time series modeling methods using deep learning.

The code for all the implementations in each chapter and the required datasets is available for download at github.com/apress/time-series-algorithm-recipes.

CHAPTER 1

Getting Started with Time Series

A *time series* is a sequence of time-dependent data points. For example, the demand (or sales) for a product in an e-commerce website can be measured temporally in a time series, where the demand (or sales) is ordered according to the time. This data can then be analyzed to find critical temporal insights and forecast future values, which helps businesses plan and increase revenue.

Time series data is used in every domain where real-time analytics is essential. Analyzing this data and forecasting its future value has become essential to these domains.

Time series analysis/forecasting was previously considered a purely statistical problem. It is now used in many machine learning and deep learning–based solutions, which perform equally well or even outperform most other solutions. This book uses various methods and approaches to analyze and forecast time series.

This chapter uses recipes to read/write time series data and perform simple preprocessing and Exploratory Data Analysis (EDA).

The following lists the recipes explored in this chapter.

Recipe 1-1. Reading Time Series Objects

Recipe 1-2. Saving Time Series Objects

© Akshay R Kulkarni, Adarsha Shivananda, Anoosh Kulkarni, V Adithya Krishnan 2023
A. R. Kulkarni et al., *Time Series Algorithms Recipes*,
https://doi.org/10.1007/978-1-4842-8978-5_1

Recipe 1-3. Exploring Types of Time Series Data

Recipe 1-4. Time Series Components

Recipe 1-5. Time Series Decomposition

Recipe 1-6. Visualization of Seasonality

Recipe 1-1A. Reading Time Series Objects (Air Passengers)

Problem

You want to read and load time series data into a dataframe.

Solution

Pandas load the data into a dataframe structure.

How It Works

The following steps read the data.

Step 1A-1. Import the required libraries.

```
import pandas as pd
import matplotlib.pyplot as plt
```

Step 1A-2. Write a parsing function for the datetime column.

Before reading the data, let's write a parsing function.

```
date_parser_fn = lambda dates: pd.datetime.strptime(dates,
'%Y-%m')
```

Step 1A-3. Read the data.

Read the air passenger data.

```
data = pd.read_csv('./data/AirPassenger.csv', parse_dates =
['Month'], index_col = 'Month', date_parser = date_parser_fn)
plt.plot(data)
plt.show()
```

Figure 1-1 shows the time series plot output.

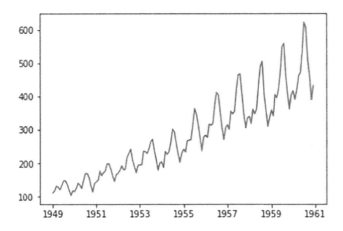

Figure 1-1. *Output*

The following are some of the important input arguments for read_csv.

- parse_dates mentions the datetime column in the dataset that needs to be parsed.

- index_col mentions the column that is a unique identifier for the pandas dataframe. In most time series use cases, it's the datetime column.

- date_parser is a function to parse the dates (i.e., converts an input string to datetime format/type). pandas reads the data in YYYY-MM-DD HH:MM:SS format. Convert to this format when using the parser function.

3

Recipe 1-1B. Reading Time Series Objects (India GDP Data)

Problem

You want to save the loaded time series dataframe in a file.

Solution

Save the dataframe as a comma-separated (CSV) file.

How It Works

The following steps read the data.

Step 1B-1. Import the required libraries.

```
import pandas as pd
import matplotlib.pyplot as plt
import pickle
```

Step 1B-2. Read India's GDP time series data.

```
indian_gdp_data = pd.read_csv('./data/GDPIndia.csv', header=0)

date_range = pd.date_range(start='1/1/1960', end='31/12/2017',
freq='A')

indian_gdp_data ['TimeIndex'] = pd.DataFrame(date_range,
columns=['Year'])
indian_gdp_data.head(5).T
```

Step 1B-3. Plot the time series.

```
plt.plot(indian_gdp_data.TimeIndex, indian_gdp_data.
GDPpercapita)
plt.legend(loc='best')
plt.show()
```

Figure 1-2 shows the output time series.

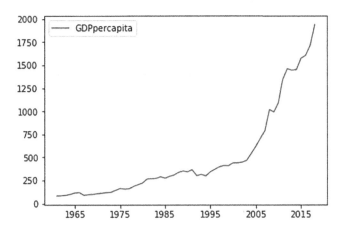

Figure 1-2. *Output*

Step 1B-4. Store and retrieve as a pickle.

```
### Store as a pickle object
import pickle
with open('gdp_india.obj', 'wb') as fp:
        pickle.dump(IndiaGDP, fp)

### Retrieve the pickle object
with open('gdp_india.obj', 'rb') as fp:
    indian_gdp_data1 = pickle.load(fp)
indian_gdp_data1.head(5).T
```

Figure 1-3 shows the retrieved time series object transposed.

	0	1	2	3	4
Year	1960	1961	1962	1963	1964
GDPpercapita	81.2848	84.4264	88.9149	100.049	114.315
TimeIndex	1960-12-31 00:00:00	1961-12-31 00:00:00	1962-12-31 00:00:00	1963-12-31 00:00:00	1964-12-31 00:00:00

Figure 1-3. *Output*

Recipe 1-2. Saving Time Series Objects

Problem

You want to save a loaded time series dataframe into a file.

Solution

Save the dataframes as a CSV file.

How It Works

The following steps store the data.

Step 2-1. Save the previously loaded time series object.

```
### Saving the TS object as csv
data.to_csv('ts_data.csv', index = True, sep = ',')

### Check the obj stored
data1 = data.from_csv('ts_data.csv', header = 0)

### Check
print(data1.head(2).T)
```

The output is as follows.

```
1981-01-01
1981-01-02    17.9
1981-01-03    18.8
Name: 20.7, dtype: float64
```

Recipe 1-3A. Exploring Types of Time Series Data: Univariate

Problem

You want to load and explore univariate time series data.

Solution

A *univariate time series* is data with a single time-dependent variable.

Let's look at a sample dataset of the monthly minimum temperatures in the Southern Hemisphere from 1981 to 1990. The temperature is the time-dependent target variable.

How It Works

The following steps read and plot the univariate data.

Step 3A-1. Import the required libraries.

```
import pandas as pd
```

```
import matplotlib.pyplot as plt
```

Step 3A-2. Read the time series data.

```
data = pd.read_csv('./data/daily-minimum-temperatures.csv',
header = 0, index_col = 0, parse_dates = True, squeeze = True)
print(data.head())
```

The output is as follows.

```
Date
1981-01-01    20.7
1981-01-02    17.9
1981-01-03    18.8
1981-01-04    14.6
1981-01-05    15.8
Name: Temp, dtype: float64
```

Step 3A-3. Plot the time series.

Let's now plot the time series data to detect patterns.

```
data.plot()
plt.ylabel('Minimum Temp')
plt.title('Min temp in Southern Hemisphere From 1981 to 1990')
plt.show()
```

Figure 1-4 shows the output time series plot.

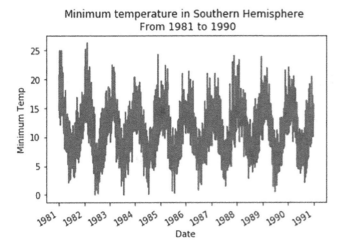

Figure 1-4. *Time series plot*

This is called *univariate time series analysis* since only one variable, temp (the temperature over the past 19 years), was used.

Recipe 1-3B. Exploring Types of Time Series Data: Multivariate

Problem

You want to load and explore multivariate time series data.

Solution

A *multivariate time series* is a type of time series data with more features that the target depends on, which are also time-dependent; that is, the target is not only dependent on its past values. This relationship is used to forecast the target values.

Let's load and explore a Beijing pollution dataset, which is multivariate.

How It Works

The following steps read and plot the multivariate data.

Step 3B-1. Import the required libraries.

```
import pandas as pd

from datetime import datetime
import matplotlib.pyplot as plt
```

Step 3B-2. Write the parsing function.

Before loading the raw dataset and parsing the datetime information as the pandas dataframe index, let's first write a parsing function.

```
def parse(x):
    return datetime.strptime(x, '%Y %m %d %H')
```

Step 3B-3. Load the dataset.

```
data1 = pd.read_csv('./data/raw.csv',  parse_dates = [['year',
'month', 'day', 'hour']],
                    index_col=0, date_parser=parse)
```

Step 3B-4. Do basic preprocessing.

Drop the No column.

```
data1.drop('No', axis=1, inplace=True)
```

Manually specify each column name.

```
data1.columns = ['pollution', 'dew', 'temp', 'press', 'wnd_
dir', 'wnd_spd', 'snow', 'rain']
data1.index.name = 'date'
```

Let's mark all NA values with 0.

```
data1['pollution'].fillna(0, inplace=True)
```

Drop the first 24 hours.

```
data1 = data1[24:]
```

Summarize the first five rows.

```
print(data1.head(5))
```

The output is as follows.

```
                     pollution  dew  temp   press wnd_dir
wnd_spd  snow  rain
date
2010-01-02 00:00:00     129.0  -16  -4.0  1020.0      SE
1.79    0    0
2010-01-02 01:00:00     148.0  -15  -4.0  1020.0      SE
2.68    0    0
2010-01-02 02:00:00     159.0  -11  -5.0  1021.0      SE
3.57    0    0
2010-01-02 03:00:00     181.0   -7  -5.0  1022.0      SE
5.36    1    0
2010-01-02 04:00:00     138.0   -7  -5.0  1022.0      SE
6.25    2    0
```

This information is from a dataset on the pollution and weather conditions in Beijing. The time aggregation of the recordings was hourly and measured for five years. The data includes the datetime column, the pollution metric known as PM2.5 concentration, and some critical weather information, including temperature, pressure, and wind speed.

Step 3B-5. Plot each series.

Now let's plot each series as a separate subplot, except wind speed direction, which is categorical.

```
vals = data1.values

# specify columns to plot

group_list = [0, 1, 2, 3, 5, 6, 7]
i = 1

# plot each column
plt.figure()

for group in group_list:
    plt.subplot(len(group_list), 1, i)
    plt.plot(vals[:, group])
    plt.title(data1.columns[group], y=0.5, loc='right')
    i += 1

plt.show()
```

Figure 1-5 shows the plot of all variables across time.

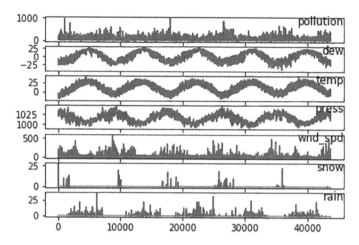

Figure 1-5. *A plot of all variables across time*

Recipe 1-4A. Time Series Components: Trends

Problem

You want to find the components of the time series, starting with trends.

Solution

A *trend* is the overall movement of data in a particular direction—that is, the values going upward (increasing) or downward (decreasing) over a period of time.

Let's use a shampoo sales dataset, which has a monthly sales count for three years.

How It Works

The following steps read and plot the data.

13

Step 4A-1. Import the required libraries.

```
import pandas as pd
import matplotlib.pyplot as plt
```

Step 4A-2. Write the parsing function.

```
def parsing_fn(x):
    return datetime.strptime('190'+x, '%Y-%m')
```

Step 4A-3. Load the dataset.

```
data = pd.read_csv('./data/shampoo-sales.csv', header=0, parse_
dates=[0], index_col=0, squeeze=True, date_parser= parsing_fn)
```

Step 4A-4. Plot the time series.

```
data.plot()
plt.show()
```

Figure 1-6 shows the time series plot.

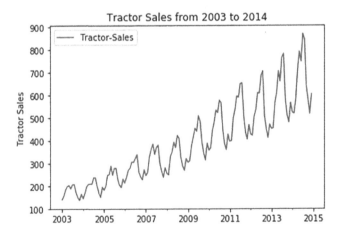

Figure 1-6. *Output*

This data has a rising trend, as seen in Figure 1-6. The output time series plot shows that, on average, the values increase with time.

Recipe 1-4B. Time Series Components: Seasonality

Problem

You want to find the components of time series data based on seasonality.

Solution

Seasonality is the recurrence of a particular pattern or change in time series data.

Let's use a Melbourne, Australia, minimum daily temperature dataset from 1981–1990. The focus is on seasonality.

How It Works

The following steps read and plot the data.

Step 4B-1. Import the required libraries.

```
import pandas as pd
import matplotlib.pyplot as plt
```

Step 4B-2. Read the data.

```
data = pd.read_csv('./data/daily-minimum-temperatures.csv',
header = 0, index_col = 0, parse_dates = True, squeeze = True)
```

Step 4B-3. Plot the time series.

```
data.plot()
plt.ylabel('Minimum Temp')
plt.title('Min temp in Southern Hemisphere from 1981 to 1990')
plt.show()
```

Figure 1-7 shows the time series plot.

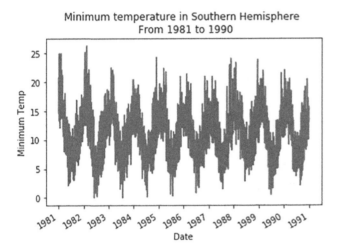

Figure 1-7. *Output*

Figure 1-7 shows that this data has a strong seasonality component (i.e., a repeating pattern in the data over time).

Step 4B-4. Plot a box plot by month.

Let's visualize a box plot to check monthly variation in 1990.

```
month_df = DataFrame()
one_year_ser = data['1990']
grouped_df = one_year_ser.groupby(Grouper(freq='M'))
month_df = pd.concat([pd.DataFrame(x[1].values) for x in
grouped_df], axis=1)
```

```
month_df = pd.DataFrame(month_df)
month_df.columns = range(1,13)
month_df.boxplot()
plt.show()
```

Figure 1-8 shows the box plot output by month.

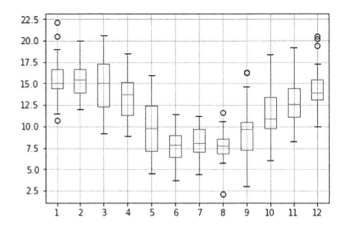

Figure 1-8. *Monthly level box plot output*

The box plot, Figure 1-8, shows the distribution of minimum temperature for each month. There appears to be a seasonal component each year, showing a swing from summer to winter. This implies a monthly seasonality.

Step 4B-5. Plot a box plot by year.

Let's group by year to see the change in distribution across various years. This way, you can check for seasonality at every time aggregation.

```
grouped_ser = data.groupby(Grouper(freq='A'))
year_df    = pd.DataFrame()
for name, group in grouped_ser:
```

```
    year_df[name.year] = group.values
year_df.boxplot()
plt.show()
```

Figure 1-9 shows the box plot output by year.

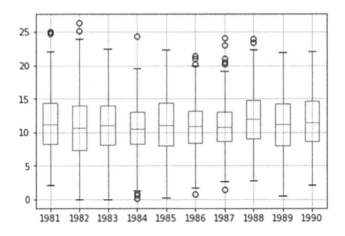

Figure 1-9. *Yearly level box plot*

Figure 1-9 reveals that there is not much yearly seasonality or trends in the box plot output.

Recipe 1-4C. Time Series Components: Seasonality (cont'd.)

Problem

You want to find time series components using another example of seasonality.

Solution

Let's explore tractor sales data to understand seasonality.

How It Works

The following steps read and plot the data.

Step 4C-1. Import the required libraries.

```
import pandas as pd
import matplotlib.pyplot as plt
```

Step 4C-2. Read tractor sales data.

```
tractor_sales_data = pd.read_csv("./data/tractor_sales
Sales.csv")
tractor_sales_data.head(5)
```

Step 4C-3. Set a datetime series to use as an index.

```
date_ser = pd.date_range(start='2003-01-01', freq='MS',
periods=len(Tractor))
```

Step 4C-4. Format the data.

```
tractor_sales_data.rename(columns={'Number of Tractor
Sold':'Tractor-Sales'}, inplace=True)
tractor_sales_data.set_index(dates, inplace=True)
tractor_sales_data = tractor_sales_data[['Tractor-Sales']]
tractor_sales_data.head(5)
```

Step 4C-5. Plot the time series.

```
tractor_sales_data.plot()
plt.ylabel('Tractor Sales')
plt.title("Tractor Sales from 2003 to 2014")
plt.show()
```

Figure 1-10 shows the time series plot output.

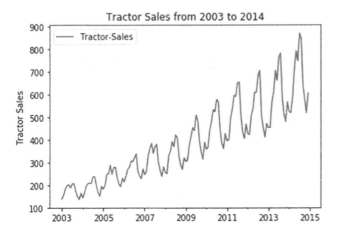

Figure 1-10. *Output*

From the time series plot, Figure 1-10 shows that the data has a strong seasonality with an increasing trend.

Step 4C-6. Plot a box plot by month.

Let's check the box plot by month to better understand the seasonality.

```
month_df = pd.DataFrame()
one_year_ser = tractor_sales_data['2011']
grouped_ser = one_year_ser.groupby(Grouper(freq='M'))
month_df = pd.concat([pd.DataFrame(x[1].values) for x in
grouped_ser], axis=1)
month_df = pd.DataFrame(month_df)
month_df.columns = range(1,13)
month_df.boxplot()
plt.show()
```

Figure 1-11 shows the box plot output by month.

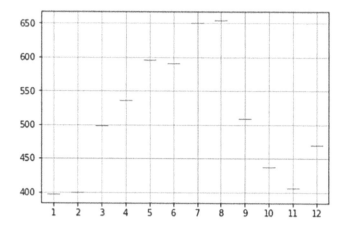

Figure 1-11. *Monthly level box plot*

The box plot shows a seasonal component each year, with a swing from May to August.

Recipe 1-5A. Time Series Decomposition: Additive Model

Problem

You want to learn how to decompose a time series using additive model decomposition.

Solution

- The additive model suggests that the components add up.

- It is linear, where changes over time are constantly made in the same amount.

- The seasonality should have the same frequency and amplitude. Frequency is the width between cycles, and amplitude is the height of each cycle.

The statsmodel library has an implementation of the classical decomposition method, but the user has to specify whether the model is additive or multiplicative. The function is called seasonal_decompose.

How It Works

The following steps load and decompose the time series.

Step 5A-1. Load the required libraries.

```
### Load required libraries
import pandas as pd
import matplotlib.pyplot as plt
from statsmodels.tsa.seasonal import seasonal_decompose
import statsmodels.api as sm
```

Step 5A-2. Read and process retail turnover data.

```
turn_over_data = pd.read_csv('./data/RetailTurnover.csv')
date_range = pd.date_range(start='1/7/1982', end='31/3/1992',
freq='Q')
turn_over_data['TimeIndex'] = pd.DataFrame(date_range,
columns=['Quarter'])
```

Step 5A-3. Plot the time series.

```
plt.plot(turn_over_data.TimeIndex, turn_over_data.Turnover)
plt.legend(loc='best')
plt.show()
```

Figure 1-12 shows the time series plot output.

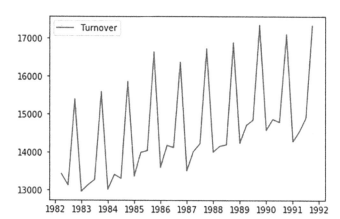

Figure 1-12. *Time series plot output*

Figure 1-12 shows that the trend is linearly increasing, and there is constant linear seasonality.

Step 5A-4. Decompose the time series.

Let's decompose the time series by trends, seasonality, and residuals.

```
decomp_turn_over = sm.tsa.seasonal_decompose(turn_over_data.
Turnover, model="additive", freq=4)
decomp_turn_over.plot()
plt.show()
```

Figure 1-13 shows the time series decomposition output.

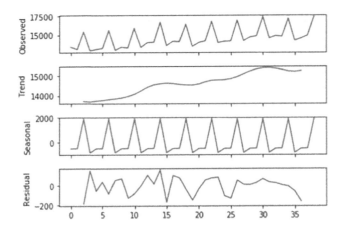

Figure 1-13. *Time series decomposition output*

Step 5A-5. Separate the components.

You can get the trends, seasonality, and residuals as separate series with the following.

```
trend = decomp_turn_over.trend
seasonal = decomp_turn_over.seasonal
residual = decomp_turn_over.resid
```

Recipe 1-5B. Time Series Decomposition: Multiplicative Model

Problem

You want to learn how to decompose a time series using multiplicative model decomposition.

Solution

- A multiplicative model suggests that the components are multiplied up.

- It is non-linear, such as quadratic or exponential, which means that the changes increase or decrease with time.

- The seasonality has an increasing or a decreasing frequency and/or amplitude.

How It Works

The following steps load and decompose the time series.

Step 5B-1. Load the required libraries.

```
### Load required libraries
import pandas as pd
import matplotlib.pyplot as plt
from statsmodels.tsa.seasonal import seasonal_decompose
import statsmodels.api as sm
```

Step 5B-2. Load air passenger data.

```
air_passengers_data = pd.read_csv('./data/AirPax.csv')
```

Step 5B-3. Process the data.

```
date_range = pd.date_range(start='1/1/1949', end='31/12/1960',
freq='M')
air_passengers_data ['TimeIndex'] = pd.DataFrame(date_range,
columns=['Month'])
print(air_passengers_data.head())
```

The output is as follows.

```
   Year Month  Pax  TimeIndex
0  1949   Jan  112  1949-01-31
1  1949   Feb  118  1949-02-28
2  1949   Mar  132  1949-03-31
3  1949   Apr  129  1949-04-30
4  1949   May  121  1949-05-31
```

Figure 1-14 shows the time series output plot.

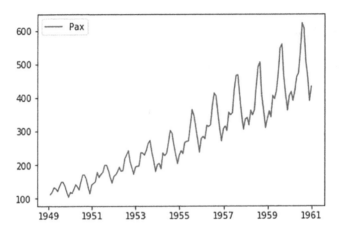

Figure 1-14.   *Time series output plot*

Step 5B-4. Decompose the time series.

```
decomp_air_passengers_data = sm.tsa.seasonal_decompose
(air_passengers_data.Pax, model="multiplicative", freq=12)
decomp_air_passengers_data.plot()
plt.show()
```

Figure 1-15 shows the time series decomposition output.

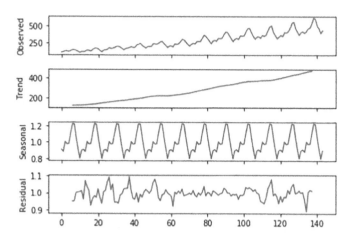

Figure 1-15. *Time series decomposition output*

Step 5B-5. Get the seasonal component.

```
Seasonal_comp = decomp_air_passengers_data.seasonal
Seasonal_comp.head(4)
The output is as follows.
```

```
0    0.910230
1    0.883625
2    1.007366
3    0.975906
Name: Pax, dtype: float64
```

Recipe 1-6. Visualization of Seasonality

Problem

You want to learn how to visualize the seasonality component.

Solution

Let's look at a few additional methods to visualize and detect seasonality. The retail turnover data shows the seasonality component per quarter.

How It Works

The following steps load and visualize the time series (i.e., the seasonality component).

Step 6-1. Import the required libraries.

```
import pandas as pd
import matplotlib.pyplot as plt
```

Step 6-2. Load the data.

```
turn_over_data = pd.read_csv('./data/RetailTurnover.csv')
```

Step 6-3. Process the data.

```
date_range = pd.date_range(start='1/7/1982', end='31/3/1992',
freq='Q')
turn_over_data['TimeIndex'] = pd.DataFrame(date_range,
columns=['Quarter'])
```

Step 6-4. Pivot the table.

Now let's pivot the table such that quarterly information is in the columns, yearly information is in the rows, and the values consist of turnover information.

```
quarterly_turn_over_data = pd.pivot_table(turn_over_data,
values = "Turnover", columns = "Quarter", index = "Year")
quarterly_turn_over_data
```

Figure 1-16 shows the output by quarterly turnover.

| Quarter | Q1 | Q2 | Q3 | Q4 |
Year				
1982	NaN	NaN	13423.2	13128.8
1983	15398.8	12964.2	13133.5	13271.7
1984	15596.3	13018.0	13409.3	13304.2
1985	15873.9	13366.5	13998.6	14045.1
1986	16650.3	13598.4	14183.2	14128.5
1987	16380.7	13512.8	14022.1	14231.8
1988	16737.0	14004.5	14165.5	14203.9
1989	16895.1	14248.2	14719.5	14855.8
1990	17361.6	14585.2	14873.5	14798.4
1991	17115.2	14284.9	14558.8	14914.3
1992	17342.3	NaN	NaN	NaN

Figure 1-16. *Quarterly turnover output*

Step 6-5. Plot the line charts.

Let's plot line plots for the four quarters.

```
quarterly_turn_over_data.plot()
plt.show()
```

Figure 1-17 shows the quarter-level line plots.

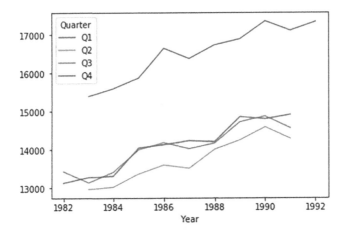

Figure 1-17. *Quarterly turnover line chart*

Step 6-6. Plot the box plots.

Let's also plot the box plot at the quarterly level.

```
quarterly_turn_over_data.boxplot()
plt.show()
```

Figure 1-18 shows the output of the box plot by quarter.

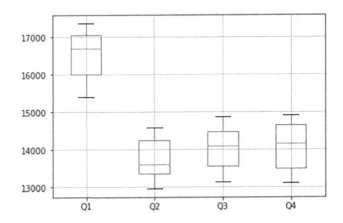

Figure 1-18. *Quarterly level box plot*

Looking at both the box plot and the line plot, you can conclude that the yearly turnover is significantly high in the first quarter and is quite low in the second quarter.

CHAPTER 2

Statistical Univariate Modeling

Univariate time series data analysis is the most popular type of temporal data, where a single numeric observation is recorded sequentially over equal time periods. Only the variable observed and its relation to time is considered in this analysis.

The forecasting of future values of this univariate data is done through univariate modeling. In this case, the predictions are dependent only on historical values. The forecasting can be done through various statistical methods. This chapter goes through a few important ones.

The following recipes for performing univariate statistical modeling are covered.

Recipe 2-1. Moving Average (MA) Forecast

Recipe 2-2. Autoregressive (AR) Model

Recipe 2-3. Autoregressive Moving Average (ARMA) Model

Recipe 2-4. Autoregressive Integrated Moving Average (ARIMA) Model

Recipe 2-5. Grid search Hyperparameter Tuning for Autoregressive Integrated Moving Average (ARIMA) Model

© Akshay R Kulkarni, Adarsha Shivananda, Anoosh Kulkarni, V Adithya Krishnan 2023
A. R. Kulkarni et al., *Time Series Algorithms Recipes*,
https://doi.org/10.1007/978-1-4842-8978-5_2

Recipe 2-6. Seasonal Autoregressive Integrated Moving Average (SARIMA) Model

Recipe 2-7. Simple Exponential Smoothing (SES) Model

Recipe 2-8. Holt-Winters (HW) Model

Recipe 2-1. Moving Average (MA) Forecast

Problem

You want to load time series data and forecast using a *moving average*.

Solution

A moving average is a method that captures the average change in a metric over time. For a particular window length, which is a short period/range in time, you calculate the mean target value, and then this window is moved across the entire period of the data, from start to end. It is usually used to smoothen the data and remove any random fluctuations.

Let's use the pandas rolling mean function to get the moving average.

How It Works

The following steps read the data and forecast using the moving average.

Step 1-1. Import the required libraries.

```
from pandas import read_csv, Grouper, DataFrame, concat
import matplotlib.pyplot as plt
from datetime import datetime
```

Step 1-2. Read the data.

The US GDP data is a time series dataset that shows the annual gross domestic product (GDP) value (in US dollars) of the United States from 1929 to 1991.

The following reads the US GDP data.

```
us_gdp_data = pd.read_csv('./data/GDPUS.csv', header=0)
```

Step 1-3. Preprocess the data.

```
date_rng = pd.date_range(start='1/1/1929', end='31/12/1991',
freq='A')
print(date_rng)

us_gdp_data['TimeIndex'] = pd.DataFrame(date_rng,
columns=['Year'])
```

The output is as follows.

```
DatetimeIndex(['1929-12-31', '1930-12-31', '1931-12-31',
'1932-12-31',
                '1933-12-31', '1934-12-31', '1935-12-31',
'1936-12-31',
                '1937-12-31', '1938-12-31', '1939-12-31',
'1940-12-31',
                '1941-12-31', '1942-12-31', '1943-12-31',
'1944-12-31',
                '1945-12-31', '1946-12-31', '1947-12-31',
'1948-12-31',
                '1949-12-31', '1950-12-31', '1951-12-31',
'1952-12-31',
                '1953-12-31', '1954-12-31', '1955-12-31',
'1956-12-31',
```

```
               '1957-12-31', '1958-12-31', '1959-12-31',
'1960-12-31',
               '1961-12-31', '1962-12-31', '1963-12-31',
'1964-12-31',
               '1965-12-31', '1966-12-31', '1967-12-31',
'1968-12-31',
               '1969-12-31', '1970-12-31', '1971-12-31',
'1972-12-31',
               '1973-12-31', '1974-12-31', '1975-12-31',
'1976-12-31',
               '1977-12-31', '1978-12-31', '1979-12-31',
'1980-12-31',
               '1981-12-31', '1982-12-31', '1983-12-31',
'1984-12-31',
               '1985-12-31', '1986-12-31', '1987-12-31',
'1988-12-31',
               '1989-12-31', '1990-12-31', '1991-12-31'],
             dtype='datetime64[ns]', freq='A-DEC')
```

Step 1-4. Plot the time series.

```
plt.plot(us_gdp_data.TimeIndex, us_gdp_data.GDP)
plt.legend(loc='best')
plt.show()
```

Figure 2-1 shows the time series plot output.

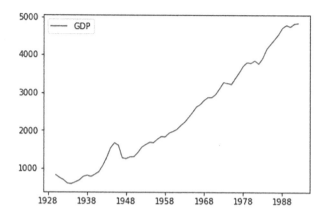

Figure 2-1. *Output*

Step 1-5. Use a rolling mean to get the moving average.

A window size of 5 is used for this example.

```
mvg_avg_us_gdp = us_gdp_data.copy()
#calculating the rolling mean - with window 5
mvg_avg_us_gdp['moving_avg_forecast'] = us_gdp_data['GDP'].
rolling(5).mean()
```

Step 1-6. Plot the forecast vs. the actual.

```
plt.plot(us_gdp_data['GDP'], label='US GDP')
plt.plot(mvg_avg_us_gdp['moving_avg_forecast'], label='US
GDP MA(5)')
plt.legend(loc='best')
plt.show()
```

Figure 2-2 shows the moving average (MA) forecast vs. the actual.

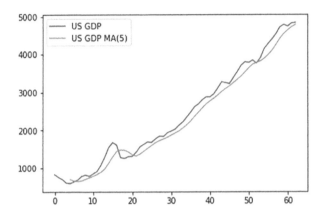

Figure 2-2. *MA forecast vs. actual*

Recipe 2-2. Autoregressive (AR) Model
Problem

You want to load the time series data and forecast using an *autoregressive model.*

Solution

Autoregressive models use lagged values (i.e., the historical values of a point to forecast future values). The forecast is a linear combination of these lagged values.

Let's use the AutoReg function from statsmodels.tsa for modeling.

How It Works

The following steps load data and forecast using the AR model.

Step 2-1. Import the required libraries.

```
import pandas as pd
import numpy as np
import matplotlib.pyplot as plt
from statsmodels.tsa.stattools import adfuller
from statsmodels.tsa.ar_model import AutoReg
from statsmodels.graphics.tsaplots import import plot_pacf
```

Step 2-2. Load and plot the dataset.

```
url='opsd_germany_daily.csv'
data = pd.read_csv(url,sep=",")
data['Consumption'].plot()
```

Figure 2-3 shows the plot of the time series data.

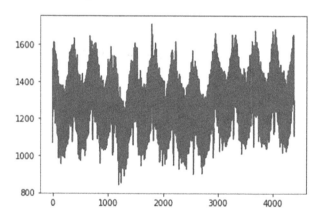

Figure 2-3. *Output*

Step 2-3. Check for stationarity of the time series data.

Let's look for the p-value in the output of the Augmented Dickey-Fuller test. If the p-value is less than 0.05, the time series is stationary.

```
data_stationarity_test = adfuller(data['Consumption'],
autolag='AIC')
print("P-value: ", data_stationarity_test[1])
```

The output is as follows.

```
P-value:   4.7440549018424435e-08
```

Step 2-4. Find the order of the AR model to be trained.

Let's plot the partial autocorrelation function (pacf) plot to assess the direct effect of past data (lags) on future data.

```
pacf = plot_pacf(data['Consumption'], lags=25)
```

Figure 2-4 shows the output of the partial autocorrelation function plot.

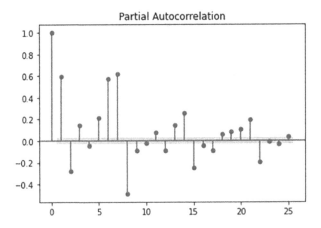

Figure 2-4. *Partial autocorrelation function plot*

Figure 2-4 shows the partial autocorrelation function output and the number of lags until there is a significant partial correlation in the order of the AR model. In this case, it is 8.

Step 2-5. Create training and test data.

```
train_df = data['Consumption'][:len(data)-100]
test_df = data['Consumption'][len(data)-100:]
```

Step 2-6. Call and fit the AR model.

```
model_ar = AutoReg(train_df, lags=8).fit()
```

Step 2-7. Output the model summary.

```
print(ar_model.summary())
```

Figure 2-5 shows the AR model summary.

```
                         AutoReg Model Results
==============================================================================
Dep. Variable:          Consumption   No. Observations:            4283
Model:                   AutoReg(8)   Log Likelihood         -24231.812
Method:            Conditional MLE    S.D. of innovations        70.058
Date:            Sat, 17 Sep 2022     AIC                         8.503
Time:                    18:12:46     BIC                         8.518
Sample:                         8     HQIC                        8.509
                             4283
==============================================================================
                   coef    std err          z      P>|z|      [0.025      0.975]
------------------------------------------------------------------------------
intercept       121.2792     14.444      8.397      0.000      92.969     149.589
Consumption.L1    0.6393      0.013     47.751      0.000       0.613       0.666
Consumption.L2   -0.0966      0.011     -8.424      0.000      -0.119      -0.074
Consumption.L3    0.0677      0.012      5.879      0.000       0.045       0.090
Consumption.L4   -0.0538      0.012     -4.653      0.000      -0.076      -0.031
Consumption.L5   -0.0092      0.012     -0.793      0.428      -0.032       0.014
Consumption.L6    0.0619      0.012      5.371      0.000       0.039       0.085
Consumption.L7    0.7832      0.011     68.283      0.000       0.761       0.806
Consumption.L8   -0.4833      0.013    -36.107      0.000      -0.510      -0.457
                                  Roots
==============================================================================
                   Real          Imaginary           Modulus         Frequency
------------------------------------------------------------------------------
AR.1            -0.9434           -0.4695j             1.0538           -0.4265
AR.2            -0.9434           +0.4695j             1.0538            0.4265
AR.3            -0.2332           -0.9929j             1.0199           -0.2867
AR.4            -0.2332           +0.9929j             1.0199            0.2867
AR.5             0.6323           -0.7958j             1.0164           -0.1431
AR.6             0.6323           +0.7958j             1.0164            0.1431
AR.7             1.0362           -0.0000j             1.0362           -0.0000
AR.8             1.6730           -0.0000j             1.6730           -0.0000
------------------------------------------------------------------------------
```

Figure 2-5. *AR model summary*

Step 2-8. Get the predictions from the model.

```
predictions = model_ar.predict(start=len(train_df),
end=(len(data)-1), dynamic=False)
```

Step 2-9. Plot the predictions vs. actuals.

```
from matplotlib import pyplot
pyplot.plot(predictions)
pyplot.plot(test_df, color='red')
```

Figure 2-6 shows the predictions vs. actuals for the AR model.

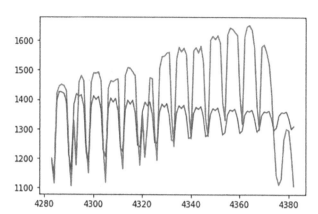

Figure 2-6. *Predictions vs. actuals*

Recipe 2-3. Autoregressive Moving Average (ARMA) Model

Problem

You want to load time series data and forecast using an *autoregressive moving average* (ARMA) model.

Solution

An ARMA model uses the concepts of autoregression and moving averages to build a much more robust model. It has two hyperparameters (p and q) that tune the autoregressive and moving average components, respectively.

Let's use the ARIMA function from statsmodels.tsa for modeling.

How It Works

The following steps load data and forecast using the ARMA model.

Step 3-1. Import the required libraries.

```
import pandas_datareader.data as web
import datetime
import pandas as pd
import numpy as np
from sklearn.metrics import mean_squared_error
import matplotlib.pyplot as plt
import seaborn as sns
from statsmodels.tsa.arima.model import ARIMA
from statsmodels.tsa.statespace.sarimax import SARIMAX
from statsmodels.tsa.api import SimpleExpSmoothing
from statsmodels.tsa.holtwinters import ExponentialSmoothing
import warnings
```

Step 3-2. Load the data.

Let's use the bitcoin price (in USD) data from December 31, 2017, to January 4, 2018.

```
btc_data = pd.read_csv("btc.csv")
print(btc_data.head())
```

The output is as follows.

```
        Date        BTC-USD
0   2017-12-31   14156.400391
1   2018-01-01   13657.200195
2   2018-01-02   14982.099609
3   2018-01-03   15201.000000
4   2018-01-04   15599.200195
```

Step 3-3. Preprocess the data.

```
btc_data.index = pd.to_datetime(btc_data['Date'],
format='%Y-%m-%d')
del btc_data['Date']
```

Step 3-4. Plot the time series.

```
plt.ylabel('Price-BTC')
plt.xlabel('Date')
plt.xticks(rotation=45)
plt.plot(btc_data.index, btc_data['BTC-USD'], )
```

Figure 2-7 shows the time series plot for the bitcoin price data.

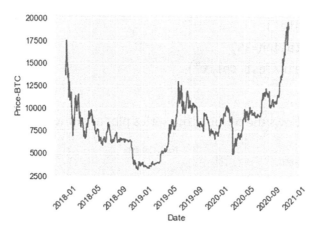

Figure 2-7. *Bitcoin price data*

Step 3-5. Do a train-test split.

```
train_data = btc_data[btc_data.index < pd.to_
datetime("2020-11-01", format='%Y-%m-%d')]
```

```
test_data = btc_data[btc_data.index > pd.to_
datetime("2020-11-01", format='%Y-%m-%d')]
print(train_data.shape)
print(test_data.shape)
```

The output is as follows.

```
(1036, 1)
(31, 1)
```

Step 3-6. Plot time the series after the train-test split.

```
plt.plot(train_data, color = "black", label = 'Train')
plt.plot(test_data, color = "green", label = 'Test')
plt.ylabel('Price-BTC')
plt.xlabel('Date')
plt.xticks(rotation=35)
plt.title("Train/Test split")
plt.show()
```

Figure 2-8 shows the output time series plot after the train-test split.

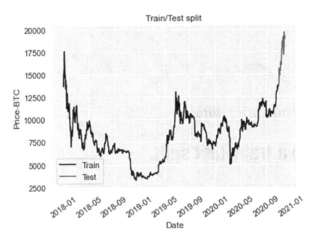

Figure 2-8. *Train-test split output*

Step 3-7. Define the actuals from training.

```
actuals = train_data['BTC-USD']
```

Step 3-8. Initialize and fit the ARMA model.

```
ARMA_model = ARIMA(actuals, order = (1, 0, 1))
ARMA_model = ARMA_model.fit()
```

Step 3-9. Get the test predictions.

```
predictions = ARMA_model.get_forecast(len(test_data.index))
predictions_df = predictions.conf_int(alpha = 0.05)
predictions_df["Predictions"] = ARMA_model.predict(start =
predictions_df.index[0], end = predictions_df.index[-1])
predictions_df.index = test_data.index
predictions_arma = predictions_df["Predictions"]
```

Step 3-10. Plot the train, test, and predictions as a line plot.

```
plt.plot(train_data, color = "black", label = 'Train')
plt.plot(test_data, color = "green", label = 'Test')
plt.ylabel('Price-BTC')
plt.xlabel('Date')
plt.xticks(rotation=35)
plt.title("ARMA model predictions")
plt.plot(predictions_arma, color="red", label = 'Predictions')
plt.legend()
plt.show()
```

Figure 2-9 shows the predictions vs. actuals for the ARMA model.

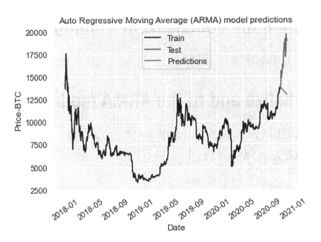

Figure 2-9. *Predictions vs. actuals output*

Step 3-11. Calculate the RMSE score for the model.

```
rmse_arma = np.sqrt(mean_squared_error(test_data["BTC-USD"].
values, predictions_df["Predictions"]))
print("RMSE: ",rmse_arma)
```

The output is as follows.

```
RMSE:   4017.145069637629
```

The RMSE (root-mean-square error) is very high, as the dataset is not stationary. You need to make it stationary or use the autoregressive integrated moving average (ARIMA) model to get a better performance.

Recipe 2-4. Autoregressive Integrated Moving Average (ARIMA) Model

Problem

You want to load time series data and forecast using an *autoregressive integrated moving average* (ARIMA) model.

Solution

An ARIMA model improves upon the ARMA model because it also includes a third parameter, d, which is responsible for differencing the data to get in stationarity for better forecasts.

Let's use the ARIMA function from statsmodels.tsa for modeling.

How It Works

The following steps load data and forecast using the ARIMA model.

Steps 3-1 to 3-7 from Recipe 2-3 are also used for this recipe.

Step 4-1. Make the data stationary by differencing.

```
# differencing
ts_diff = actuals - actuals.shift(periods=4)
ts_diff.dropna(inplace=True)
```

Step 4-2. Check the ADF (Augmented Dickey-Fuller) test for stationarity.

```
# checking for stationarity
from statsmodels.tsa.stattools import adfuller
result = adfuller(ts_diff)
```

```
pval = result[1]
print('ADF Statistic: %f' % result[0])
print('p-value: %f' % result[1])
```

The output is as follows.

```
ADF Statistic: -6.124168
p-value: 0.000000
```

Step 4-3. Get the Auto Correlation Function and Partial Auto Correlation Function values.

```
from statsmodels.tsa.stattools import adfuller
lag_acf = acf(ts_diff, nlags=20)
lag_pacf = pacf(ts_diff, nlags=20, method='ols')
```

Step 4-4. Plot the ACF and PACF to get p- and q-values.

Plot the ACF and PACF to get the q- and p-values, respectively.

```
#Ploting ACF:
plt.figure(figsize = (15,5))
plt.subplot(121)
plt.stem(lag_acf)
plt.axhline(y = 0, linestyle='--',color='black')
plt.axhline(y = -1.96/np.sqrt(len(ts_diff)),linestyle='--
',color='gray')
plt.axhline(y = 1.96/np.sqrt(len(ts_diff)),linestyle='--
',color='gray')
plt.xticks(range(0,22,1))
plt.xlabel('Lag')
plt.ylabel('ACF')
plt.title('Autocorrelation Function')

#Plotting PACF:
plt.subplot(122)
```

```
plt.stem(lag_pacf)
plt.axhline(y = 0, linestyle = '--', color = 'black')
plt.axhline(y =-1.96/np.sqrt(len(actuals)), linestyle = '--',
color = 'gray')
plt.axhline(y = 1.96/np.sqrt(len(actuals)),linestyle = '--',
color = 'gray')
plt.xlabel('Lag')
plt.xticks(range(0,22,1))
plt.ylabel('PACF')
plt.title('Partial Autocorrelation Function')
plt.tight_layout()
plt.show()
```

Figure 2-10 shows the ACF and PACF plot outputs.

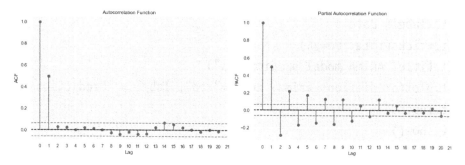

Figure 2-10. *ACF and PACF plots*

According to the ACF plot, the cutoff is 1, so the q-value is 1. According to the PACF plot, the cutoff is 10, so the p-value is 10.

Step 4-5. Initialize and fit the ARIMA model.

Using the derived p-, d-, and q-values.

```
ARIMA_model = ARIMA(actuals, order = (10, 4, 1))
ARIMA_model = ARIMA_model.fit()
```

Step 4-6. Get the test predictions.

```
predictions = ARIMA_model.get_forecast(len(test_data.index))
predictions_df = predictions.conf_int(alpha = 0.05)
predictions_df["Predictions"] = ARIMA_model.predict(start =
predictions_df.index[0], end = predictions_df.index[-1])
predictions_df.index = test_data.index
predictions_arima = predictions_df["Predictions"]
```

Step 4-7. Plot the train, test, and predictions as a line plot.

```
plt.plot(train_data, color = "black", label = 'Train')
plt.plot(test_data, color = "green", label = 'Test')
plt.ylabel('Price-BTC')
plt.xlabel('Date')
plt.xticks(rotation=35)
plt.title("ARIMA model predictions")
plt.plot(predictions_arima, color="red", label = 'Predictions')
plt.legend()
plt.show()
```

Figure 2-11 shows the predictions vs. actuals for the ARIMA model.

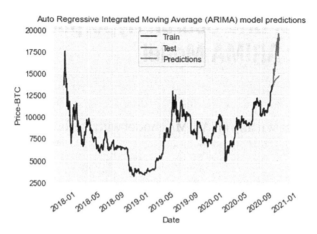

Figure 2-11. *Predictions vs. actuals output*

Step 4-8. Calculate the RMSE score for the model.

```
rmse_arima = np.sqrt(mean_squared_error(test_data["BTC-USD"].
values, predictions_df["Predictions"]))
print("RMSE: ",rmse_arima)
```

The output is as follows.

```
RMSE:  2895.312718157126
```

This model has performed better than an ARMA model due to the
differencing part and finding the proper p-, d-, and q-values. But still, it has
a high RMSE as the model is not perfectly tuned.

Recipe 2-5. Grid Search Hyperparameter Tuning for ARIMA Model

Problem

You want to forecast using an ARIMA model with the best hyperparameters.

Solution

Let's use a grid search method to tune the model's hyperparameters. The ARIMA model has three parameters (p, d, and q) that can be tuned using the classical grid search method. Loop through various combinations and evaluate each model to find the best configuration.

How It Works

The following steps load data and tune hyperparameters before forecasting using the ARIMA model.

Steps 3-1 to 3-7 from Recipe 2-3 are also used for this recipe.

Step 5-1. Write a function to evaluate the ARIMA model.

This function returns the RMSE score for a particular ARIMA order (input). It performs the same task as steps 3-8 and 3-9 in Recipe 2-3.

```
def arima_model_evaluate(train_actuals, test_data, order):
    # Model initalize and fit
    ARIMA_model = ARIMA(actuals, order = order)
    ARIMA_model = ARIMA_model.fit()
    # Getting the predictions
```

```
predictions = ARIMA_model.get_forecast(len(test_
data.index))
predictions_df = predictions.conf_int(alpha = 0.05)
predictions_df["Predictions"] = ARIMA_model.predict(start =
predictions_df.index[0], end = predictions_df.index[-1])
predictions_df.index = test_data.index
predictions_arima = predictions_df["Predictions"]
# calculate RMSE score
rmse_score = np.sqrt(mean_squared_error(test_data["BTC-
USD"].values, predictions_df["Predictions"]))
return rmse_score
```

Step 5-2. Write a function to evaluate multiple models through grid search hyperparameter tuning.

This function uses the arima_model_evaluate function defined in step 5-8 to calculate the RMSE scores of multiple ARIMA models and returns the best model/configuration. It takes as input the list of p-, d-, and q-values that needs to be tested/experimented with.

```
def evaluate_models(train_actuals, test_data, list_p_values,
list_d_values, list_q_values):
    best_rmse, best_config = float("inf"), None
    for p in list_p_values:
        for d in list_d_values:
            for q in list_q_values:
                arima_order = (p,d,q)
                rmse = arima_model_evaluate(train_actuals,
                test_data, arima_order)
                if rmse < best_rmse:
                    best_rmse, best_config = rmse, arima_order
                print('ARIMA%s RMSE=%.3f' % (arima_order,rmse))
```

```
print('Best Configuration: ARIMA%s , RMSE=%.3f' % (best_
config, best_rmse))
return best_config
```

Step 5-3. Perform the grid search hyperparameter tuning by calling the defined functions.

```
p_values = range(0, 4)
d_values = range(0, 4)
q_values = range(0, 4)
warnings.filterwarnings("ignore")
best_config = evaluate_models(actuals, test_data, p_values,
d_values, q_values)
```

The output is as follows.

```
ARIMA(0, 0, 0) RMSE=8973.268
ARIMA(0, 0, 1) RMSE=8927.094
ARIMA(0, 0, 2) RMSE=8895.924
ARIMA(0, 0, 3) RMSE=8861.499
ARIMA(0, 1, 0) RMSE=3527.133
ARIMA(0, 1, 1) RMSE=3537.297
ARIMA(0, 1, 2) RMSE=3519.475
ARIMA(0, 1, 3) RMSE=3514.476
ARIMA(0, 2, 0) RMSE=1112.565
ARIMA(0, 2, 1) RMSE=3455.709
ARIMA(0, 2, 2) RMSE=3315.249
ARIMA(0, 2, 3) RMSE=3337.231
ARIMA(0, 3, 0) RMSE=30160.941
ARIMA(0, 3, 1) RMSE=887.423
ARIMA(0, 3, 2) RMSE=3209.141
ARIMA(0, 3, 3) RMSE=2970.229
ARIMA(1, 0, 0) RMSE=4079.516
```

```
ARIMA(1, 0, 1) RMSE=4017.145
ARIMA(1, 0, 2) RMSE=4065.809
ARIMA(1, 0, 3) RMSE=4087.934
ARIMA(1, 1, 0) RMSE=3537.539
ARIMA(1, 1, 1) RMSE=3535.791
ARIMA(1, 1, 2) RMSE=3537.341
ARIMA(1, 1, 3) RMSE=3504.703
ARIMA(1, 2, 0) RMSE=725.218
ARIMA(1, 2, 1) RMSE=3318.935
ARIMA(1, 2, 2) RMSE=3507.106
ARIMA(1, 2, 3) RMSE=3314.726
ARIMA(1, 3, 0) RMSE=12360.360
ARIMA(1, 3, 1) RMSE=727.351
ARIMA(1, 3, 2) RMSE=2968.820
ARIMA(1, 3, 3) RMSE=3019.434
ARIMA(2, 0, 0) RMSE=4014.318
ARIMA(2, 0, 1) RMSE=4022.540
ARIMA(2, 0, 2) RMSE=4062.346
ARIMA(2, 0, 3) RMSE=4088.628
ARIMA(2, 1, 0) RMSE=3522.798
ARIMA(2, 1, 1) RMSE=3509.829
ARIMA(2, 1, 2) RMSE=3523.407
ARIMA(2, 1, 3) RMSE=3517.972
ARIMA(2, 2, 0) RMSE=748.267
ARIMA(2, 2, 1) RMSE=3498.685
ARIMA(2, 2, 2) RMSE=3514.870
ARIMA(2, 2, 3) RMSE=3310.798
ARIMA(2, 3, 0) RMSE=33486.993
ARIMA(2, 3, 1) RMSE=797.942
ARIMA(2, 3, 2) RMSE=2979.751
ARIMA(2, 3, 3) RMSE=2965.450
```

```
ARIMA(3, 0, 0) RMSE=4060.745
ARIMA(3, 0, 1) RMSE=4114.216
ARIMA(3, 0, 2) RMSE=4060.737
ARIMA(3, 0, 3) RMSE=3810.374
ARIMA(3, 1, 0) RMSE=3509.046
ARIMA(3, 1, 1) RMSE=3499.516
ARIMA(3, 1, 2) RMSE=3520.499
ARIMA(3, 1, 3) RMSE=3521.356
ARIMA(3, 2, 0) RMSE=1333.102
ARIMA(3, 2, 1) RMSE=3482.502
ARIMA(3, 2, 2) RMSE=3451.985
ARIMA(3, 2, 3) RMSE=3285.008
ARIMA(3, 3, 0) RMSE=14358.749
ARIMA(3, 3, 1) RMSE=1477.509
ARIMA(3, 3, 2) RMSE=3142.936
ARIMA(3, 3, 3) RMSE=2957.573
Best Configuration: ARIMA(1, 2, 0) , RMSE=725.218
```

Step 5-4. Initialize and fit the ARIMA model with the best configuration.

```
ARIMA_model = ARIMA(actuals, order = best_config)
ARIMA_model = ARIMA_model.fit()
```

Step 5-5. Get the test predictions.

```
predictions = ARIMA_model.get_forecast(len(test_data.index))
predictions_df = predictions.conf_int(alpha = 0.05)
predictions_df["Predictions"] = ARIMA_model.predict(start =
predictions_df.index[0], end = predictions_df.index[-1])
predictions_df.index = test_data.index
predictions_arima = predictions_df["Predictions"]
```

Step 5-6. Plot the train, test, and predictions as a line plot.

```
plt.plot(train_data, color = "black", label = 'Train')
plt.plot(test_data, color = "green", label = 'Test')
plt.ylabel('Price-BTC')
plt.xlabel('Date')
plt.xticks(rotation=35)
plt.title("ARIMA model predictions")
plt.plot(predictions_arima, color="red", label = 'Predictions')
plt.legend()
plt.show()
```

Figure 2-12 shows the predictions vs. actuals for the ARIMA model.

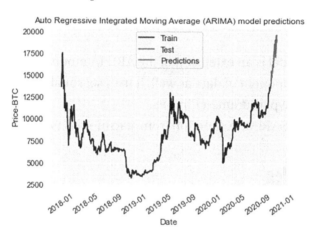

Figure 2-12. *Predictions vs. actuals output*

Step 5-7. Calculate the RMSE score for the model.

```
rmse_arima = np.sqrt(mean_squared_error(test_data["BTC-USD"].
values, predictions_df["Predictions"]))
print("RMSE: ",rmse_arima)
```

The output is as follows.

```
RMSE:   725.2180143501593
```

This is the best RMSE so far because the model is tuned and fits well.

Recipe 2-6. Seasonal Autoregressive Integrated Moving Average (SARIMA) Model

Problem

You want to load time series data and forecast using a *seasonal autoregressive integrated moving average* (SARIMA) model.

Solution

The SARIMA model is an extension of the ARIMA model. It can model the seasonal component of data as well. It uses seasonal p, d, and q components as hyperparameter inputs.

Let's use the SARIMAX function from statsmodels.tsa for modeling.

How It Works

The following steps load data and forecast using the SARIMA model.

Steps 3-1 to 3-7 from Recipe 2-3 are also used for this recipe.

Step 6-1. Initialize and fit the SARIMA model.

```
SARIMA_model = SARIMAX(actuals, order = (1, 2, 0), seasonal_
order=(2,2,2,12))
SARIMA_model = SARIMA_model.fit()
```

Step 6-2. Get the test predictions.

```
predictions = SARIMA_model.get_forecast(len(test_data.index))
predictions_df = predictions.conf_int(alpha = 0.05)
predictions_df["Predictions"] = SARIMA_model.predict(start =
predictions_df.index[0], end = predictions_df.index[-1])
predictions_df.index = test_data.index
predictions_sarima = predictions_df["Predictions"]
```

Step 6-3. Plot the train, test, and predictions as a line plot.

```
plt.plot(train_data, color = "black", label = 'Train')
plt.plot(test_data, color = "green", label = 'Test')
plt.ylabel('Price-BTC')
plt.xlabel('Date')
plt.xticks(rotation=35)
plt.title("SARIMA model predictions")
plt.plot(predictions_sarima, color="red", label =
'Predictions')
plt.legend()
plt.show()
```

Figure 2-13 shows the predictions vs. actuals for the seasonal ARIMA model.

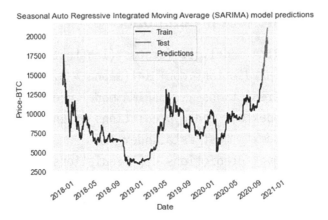

Figure 2-13. *Predictions vs. actuals output*

Step 6-4. Calculate the RMSE score for the model.

```
rmse_sarima = np.sqrt(mean_squared_error(test_data["BTC-USD"].
values, predictions_df["Predictions"]))
print("RMSE: ",rmse_sarima)
```

The output is as follows.

```
RMSE:   1050.157033576061
```

You can further tune the seasonal component to get a better RMSE score. Tuning can be done using the same grid search method.

Recipe 2-7. Simple Exponential Smoothing (SES) Model

Problem

You want to load the time series data and forecast using a *simple exponential smoothing* (SES) model.

Solution

Simple exponential smoothing is a smoothening method (like moving average) that uses an exponential window function.

Let's use the SimpleExpSmoothing function from statsmodels.tsa for modeling.

How It Works

The following steps load data and forecast using the SES model.

Steps 3-1 to 3-7 from Recipe 2-3 are also used for this recipe.

Step 7-1. Initialize and fit the SES model.

```
SES_model = SimpleExpSmoothing(actuals)
SES_model = SES_model.fit(smoothing_level=0.8,optimized=False)
```

Step 7-2. Get the test predictions.

```
predictions_ses = SES_model.forecast(len(test_data.index))
```

Step 7-3. Plot the train, test, and predictions as a line plot.

```
plt.plot(train_data, color = "black", label = 'Train')
plt.plot(test_data, color = "green", label = 'Test')
plt.ylabel('Price-BTC')
plt.xlabel('Date')
plt.xticks(rotation=35)
plt.title("SImple Exponential Smoothing (SES) model
predictions")
plt.plot(predictions_ses, color='red', label = 'Predictions')
plt.legend()
plt.show()
```

Figure 2-14 shows the predictions vs. actuals for the SES model.

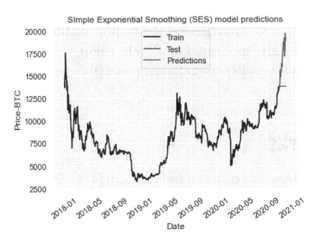

Figure 2-14. *Predictions vs. actuals output*

Step 7-4. Calculate the RMSE score for the model.

```
rmse_ses = np.sqrt(mean_squared_error(test_data["BTC-USD"].
values, predictions_ses))
print("RMSE: ",rmse_ses)
```

The output is as follows.

```
RMSE:   3536.5763879303104
```

As expected, the RMSE is very high because it's a simple smoothing function that performs best when there is no trend in the data.

Recipe 2-8. Holt-Winters (HW) Model Problem

You want to load time series data and forecast using the Holt-Winters (HW) model.

Solution

Holt-Winters is also a smoothing function. It uses the exponential weighted moving average. It encodes previous historical values to predict present and future values.

For modeling, let's use the ExponentialSmoothing function from statsmodels.tsa.holtwinters.

How It Works

The following steps load data and forecast using the HW model.

Steps 3-1 to 3-7 from Recipe 2-3 are also used for this recipe.

Step 8-1. Initialize and fit the HW model.

```
HW_model = ExponentialSmoothing(actuals, trend='add')
HW_model = HW_model.fit()
```

Step 8-2. Get the test predictions.

```
predictions_hw = HW_model.forecast(len(test_data.index))
```

Step 8-3. Plot the train, test, and predictions as a line plot.

```
plt.plot(train_data, color = "black", label = 'Train')
plt.plot(test_data, color = "green", label = 'Test')
plt.ylabel('Price-BTC')
plt.xlabel('Date')
plt.xticks(rotation=35)
plt.title("HW model predictions")
plt.plot(predictions_hw, color='red', label = 'Predictions')
plt.legend()
plt.show()
```

Figure 2-15 shows the predictions vs. actuals for the HW model.

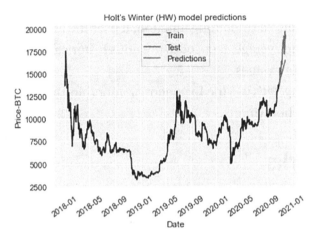

Figure 2-15. *Predictions vs. actuals output*

Step 8-4. Calculate the RMSE score for the model.

```
rmse_hw = np.sqrt(mean_squared_error(test_data["BTC-USD"].
values, predictions_hw))
print("RMSE: ",rmse_hw)
```

The output is as follows.

```
RMSE:   2024.6833967531811
```

The RMSE is a bit high, but for this dataset, the additive model performs better than the multiplicative model. For the multiplicative model, change the trend term to 'mul' in the ExponentialSmoothing function.

CHAPTER 3

Advanced Univariate and Statistical Multivariate Modeling

Chapter 2 explored various recipes for implementing univariate statistical modeling in Python. A few more advanced techniques are explored in this chapter, as well as modeling another type of temporal data—the multivariate time series. Multivariate time series contains additional time-dependent features that impact your target, apart from the date and time. The various statistical methods and recipes for implementing multivariate modeling in Python are explored.

This chapter covers the following recipes for performing advanced univariate and statistical multivariate modeling.

Recipe 3-1. FBProphet Univariate Time Series Modeling

Recipe 3-2. FBProphet Modeling by Controlling the Change Points

Recipe 3-3. FBProphet Modeling by Adjusting Trends

Recipe 3-4. FBProphet Modeling with Holidays

© Akshay R Kulkarni, Adarsha Shivananda, Anoosh Kulkarni, V Adithya Krishnan 2023
A. R. Kulkarni et al., *Time Series Algorithms Recipes*,
https://doi.org/10.1007/978-1-4842-8978-5_3

Recipe 3-5. FBProphet Modeling with Added
Regressors

Recipe 3-6. Time Series Forecasting Using
Auto-ARIMA

Recipe 3-7. Multivariate Time Series Forecasting
Using the VAR Model

Recipe 3-1. FBProphet Univariate Time Series Modeling

Problem

You want to load the time series data and forecast using the Facebook
Prophet model.

Solution

Facebook's Prophet algorithm was released in 2017. It has been a game-
changer in univariate time series modeling. This algorithm performs very
well on data with additive trends and multiple seasonalities. It has been
widely used in making accurate forecasts in various domains.

How It Works

The following steps read the data and forecast using FBProphet.

Step 1-1. Import the required libraries.

```
import numpy as np
import pandas as pd
from fbprophet import Prophet
```

```
from fbprophet.plot import plot_plotly, add_
changepoints_to_plot
from sklearn.model_selection import train_test_split
import plotly.offline as py
import matplotlib.pyplot as plt
py.init_notebook_mode()
%matplotlib inline
```

Step 1-2. Read the data.

The following reads the avocado dataset.

```
df = pd.read_csv("avocado.csv").drop(columns=["Unnamed: 0"])
df.head()
```

Figure 3-1 shows the first rows of the avocado dataset.

	Date	AveragePrice	Total Volume	4046	4225	4770	Total Bags	Small Bags	Large Bags	XLarge Bags	type	year	region
0	2015-12-27	1.33	64236.62	1036.74	54454.85	48.16	8696.87	8603.62	93.25	0.0	conventional	2015	Albany
1	2015-12-20	1.35	54876.98	674.28	44638.81	58.33	9505.56	9408.07	97.49	0.0	conventional	2015	Albany
2	2015-12-13	0.93	118220.22	794.70	109149.67	130.50	8145.35	8042.21	103.14	0.0	conventional	2015	Albany
3	2015-12-06	1.08	78992.15	1132.00	71976.41	72.58	5811.16	5677.40	133.76	0.0	conventional	2015	Albany
4	2015-11-29	1.28	51039.60	941.48	43838.39	75.78	6183.95	5986.26	197.69	0.0	conventional	2015	Albany

Figure 3-1. *Avocado dataset head*

Step 1-3. Create the training dataset.

```
train_df = pd.DataFrame()
train_df['ds'] = pd.to_datetime(df["Date"])
train_df['y'] = df.iloc[:,1]
train_df.head()
```

Figure 3-2 shows the first rows of the training dataframe.

	ds	y
0	2015-12-27	1.33
1	2015-12-20	1.35
2	2015-12-13	0.93
3	2015-12-06	1.08
4	2015-11-29	1.28

Figure 3-2. *Training the dataframe head*

Step 1-4. Initialize a basic Facebook Prophet model.

```
# Initializing basic prophet model:
basic_prophet_model = Prophet()
basic_prophet_model.fit(train_df)
```

Step 1-5. Create the future dataframe for forecasting.

```
future_df = basic_prophet_model.make_future_
dataframe(periods=300)
future_df.tail()
```

Figure 3-3 shows the tail of the future dataframe.

	ds
464	2019-01-15
465	2019-01-16
466	2019-01-17
467	2019-01-18
468	2019-01-19

Figure 3-3. *Future dataframe tail*

Step 1-6. Getting the predictions.

```
# Getting the forecast
forecast_df = basic_prophet_model.predict(future_df)
```

Step 1-7. Plot the forecast.

```
plot1 = basic_prophet_model.plot(forecast_df)
```

Figure 3-4 shows the forecast output.

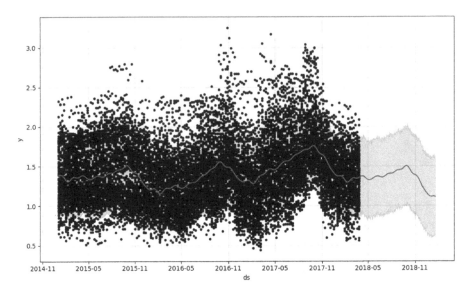

Figure 3-4. *Forecast output*

Figure 3-4 shows the blue line, which is the forecasted value.

Step 1-8. Plot the forecast components.

```
# to view the forecast components
plot2 = basic_prophet_model.plot_components(forecast_df)
```

Figure 3-5 shows the trend and seasonality components.

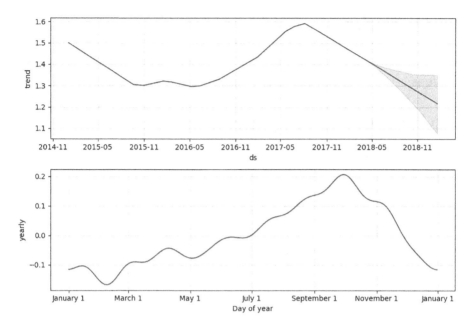

Figure 3-5. *Components output*

Recipe 3-2. FBProphet Modeling by Controlling the Change Points

Problem

You want to forecast using the Facebook Prophet model and tweak the change points.

Solution

Change points are the exact points in a time series where abrupt changes are detected. By default, the number of change points is set to 25 in the initial 80% of the time series. Let's tweak these default values by adjusting the n_change points and changepoint_range parameters to control the forecast output.

73

How It Works

The following steps forecast using FBProphet and tweak the change points.

Steps 1-1 to 1-6 from Recipe 3-1 are also used for this recipe.

Step 2-1. Plot the change points.

Let's plot the change points for the basic Prophet model in the forecast plot.

```
plot3 = basic_prophet_model.plot(forecast_df)
adding_changepoints = add_changepoints_to_plot(plot3.gca(),
basic_prophet_model, forecast_df)
```

Figure 3-6 shows the change points plotted on the forecast plot.

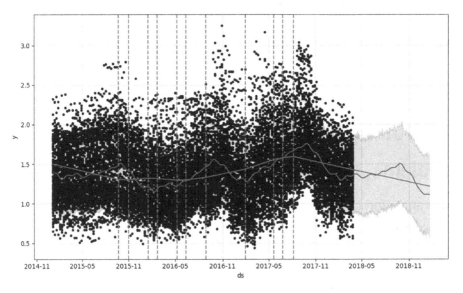

Figure 3-6. *Forecast plot with change points*

In Figure 3-6, the red dotted lines represent the change points detected by FBProphet.

Step 2-2. Print the change points.

```
basic_prophet_model.changepoints
```

The output is as follows.

```
584      2015-02-08
1168     2015-03-15
1752     2015-04-26
2336     2015-05-31
2920     2015-07-12
3504     2015-08-16
4087     2015-09-20
4671     2015-11-01
5255     2015-12-06
5839     2016-01-17
6423     2016-02-21
7007     2016-03-27
7591     2016-05-08
8175     2016-06-12
8759     2016-07-24
9343     2016-08-28
9927     2016-10-02
10511    2016-11-13
11094    2016-12-18
11678    2017-01-29
12262    2017-03-05
12846    2017-04-09
13430    2017-05-21
14014    2017-06-25
14598    2017-08-06
Name: ds, dtype: datetime64[ns]
```

Step 2-3. Check the magnitude of each changepoint.

```
deltas = basic_prophet_model.params['delta'].mean(0)
plot4 = plt.figure(facecolor='w')
ax = plot4.add_subplot(111)
ax.bar(range(len(deltas)), deltas)
ax.grid(True, which='major', c='gray', ls='-', lw=1, alpha=0.2)
ax.set_ylabel('Rate change')
ax.set_xlabel('Potential changepoint')
plot4.tight_layout()
```

Figure 3-7 shows the magnitude of each change point.

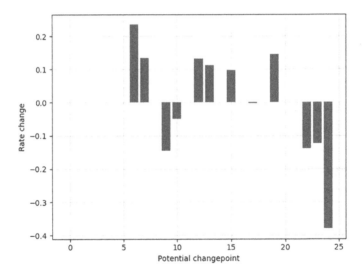

Figure 3-7. *Change points magnitude*

Step 2-4. Tweak the n_changepoints hyperparameter and forecasting.

```
# setting the n_changepoints as hyperparameter:
prophet_model_changepoint = Prophet(n_changepoints=20, yearly_
seasonality=True)
```

```
# getting the forecast
forecast_df_changepoint = prophet_model_changepoint.fit(train_
df).predict(future_df)
# plotting the forecast with change points
plot5 = prophet_model_changepoint.plot(forecast_df_changepoint)
adding_changepoints = add_changepoints_to_plot(plot5.gca(),
prophet_model_changepoint, forecast_df_changepoint)
```

Figure 3-8 shows the forecast output.

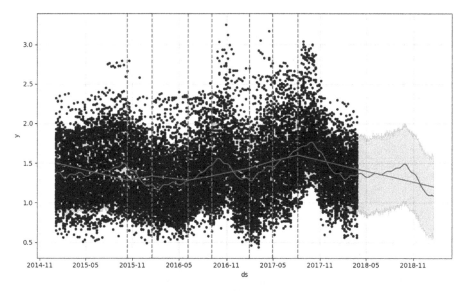

Figure 3-8. *Forecast output*

One can control the number of change points (increase or decrease) and tweak the forecast output accordingly. Greatly increasing the change points might lead to overfitting, and greatly decreasing them leads to underfitting.

Step 2-5. Tweak the changepoint_range hyperparameter and forecasting.

```
# setting the changepoint_range as hyperparameter:
prophet_model_changepoint2 = Prophet(changepoint_range=0.9,
yearly_seasonality=True)
# getting the forecast
forecast_df_changepoint2 = prophet_model_changepoint2.
fit(train_df).predict(future_df)
# plotting the forecast with change points
plot6 = prophet_model_changepoint2.plot(forecast_df_
changepoint2)
adding_changepoints = add_changepoints_to_plot(plot5.gca(),
prophet_model_changepoint2, forecast_df_changepoint2)
```

Figure 3-9 shows the forecast output.

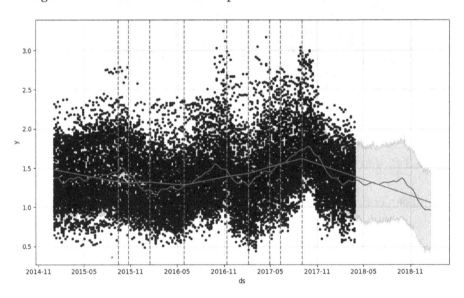

Figure 3-9. *Forecast output*

You can control the spread of the change points' location in the time series by tweaking the changepoint_range parameter. It is a percentage value and indicates the portion of the time series (from the start) where the change points are located.

Recipe 3-3. FBProphet Modeling by Adjusting Trends

Problem

You want to forecast using the Facebook Prophet model and tweak the trend component.

Solution

changepoint_prior_scale resolves the overfitting or underfitting of the model. By tweaking its value, the trend complexity changes. Increasing it increases the complexity, and decreasing it decreases complexity.

How It Works

The following steps forecast using FBProphet and tweak the trend-related parameter (i.e., changepoint_prior_scale).

Steps 1-1 to 1-3 from Recipe 3-1 are also used for this recipe.

Step 3-1. Increase the changepoint_prior_scale hyperparameter.

```
prophet_model_trend = Prophet(n_changepoints=20, yearly_
seasonality=True, changepoint_prior_scale=0.08)
```

Step 3-2. Forecast and plot the output.

```
#getting the forecast
forecast_df_trend = prophet_model_trend.fit(train_df).
predict(future_df)
# plotting the forecast with change points
plot7 = prophet_model_trend.plot(forecast_df_trend)
adding_changepoints = add_changepoints_to_plot(plot7.gca(),
prophet_model_trend, forecast_df_trend)
```

Figure 3-10 shows the forecast output.

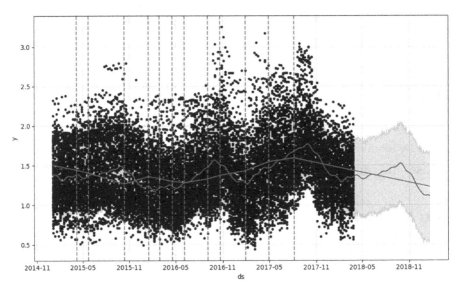

Figure 3-10.   *Forecast output*

Figure 3-10 illustrates that increasing changepoint_prior_scale increases the trend complexity. Increasing it too much can lead to overfitting, however.

Step 3-3. Decrease the changepoint_prior_scale hyperparameter.

```
prophet_model_trend2 = Prophet(n_changepoints=20, yearly_
seasonality=True, changepoint_prior_scale=0.001)
```

Step 3-4. Forecast and plot the output.

```
# getting the forecast
forecast_df_trend2 = prophet_model_trend2.fit(train_df).
predict(future_df)
# plotting the forecast with change points
plot8 = prophet_model_trend2.plot(forecast_df_trend2)
adding_changepoints = add_changepoints_to_plot(plot8.gca(),
prophet_model_trend2, forecast_df_trend2)
```

Figure 3-11 shows the forecast output.

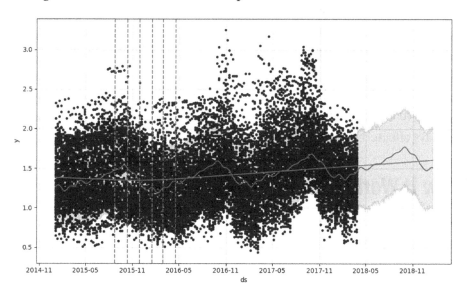

Figure 3-11. *Forecast output*

Figure 3-11 shows that decreasing the changepoint_prior_scale decreases the trend complexity. Decreasing it too much can lead to underfitting, though.

Recipe 3-4. FBProphet Modeling with Holidays

Problem

You want to forecast using the Facebook Prophet model while considering holiday data.

Solution

Information on holidays is another input parameter that can be added to the Prophet model, which uses this information when considering changes in the time series. During the holidays, it is expected that the time series will experience significant changes; hence, the model does not learn this pattern.

In the holiday dataframe, there are a series of days that are holidays. Upper and lower windows can also be provided, indicating an extension of the days effected before and after a holiday.

How It Works

The following steps forecast using FBProphet and add the holidays component.

Steps 1-1 to 1-3 from Recipe 3-1 are also used for this recipe.

Step 4-1. Create a custom holiday dataframe.

```
holidays_df = pd.DataFrame({
  'holiday': 'avocado season',
  'ds': pd.to_datetime(['2014-07-31', '2014-09-16',
                        '2015-07-31', '2015-09-16',
                        '2016-07-31', '2016-09-16',
                        '2017-07-31', '2017-09-16',
                        '2018-07-31', '2018-09-16',
                        '2019-07-31', '2019-09-16']),
  'lower_window': -1,
  'upper_window': 0,
})
```

Step 4-2. Initialize and fit the Prophet model with the holidays dataframe.

```
prophet_model_holiday = Prophet(holidays=holidays_df)
prophet_model_holiday.fit(train_df)
```

Step 4-3. Create a future dataframe for the forecast.

```
future_df = prophet_model_holiday.make_future_
dataframe(periods=12, freq = 'm')
```

Step 4-4. Get the forecast.

```
forecast_df = prophet_model_holiday.predict(future_df)
prophet_model_holiday.plot(forecast_df)
```

Figure 3-12 shows the forecast output.

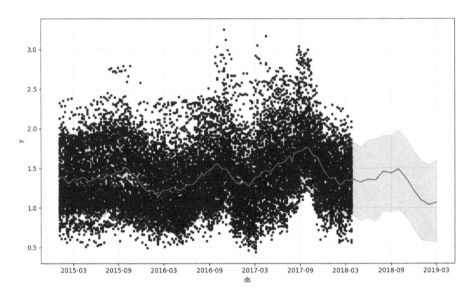

Figure 3-12. *Forecast output*

Recipe 3-5. FBProphet Modeling with Added Regressors

Problem

You want to forecast using the Facebook Prophet model with added regressors.

Solution

Additional regressors can be added to the model by using the add_ regressor functionality.

How It Works

The following steps forecast using FBProphet with added regressors.

Steps 1-1 and 1-2 from Recipe 3-1 are also used for this recipe. '

Step 5-1. Label and encode the type column.

```
# Label encoding type column
from sklearn.preprocessing import LabelEncoder
le = LabelEncoder()
df.iloc[:,10] = le.fit_transform(df.iloc[:,10])
df.head(2)
```

Figure 3-13 shows the output dataframe.

	Date	AveragePrice	Total Volume	4046	4225	4770	Total Bags	Small Bags	Large Bags	XLarge Bags	type	year	region
0	2015-12-27	1.33	64236.62	1036.74	54454.85	48.16	8696.87	8603.62	93.25	0.0	0	2015	Albany
1	2015-12-20	1.35	54876.98	674.28	44638.81	58.33	9505.56	9408.07	97.49	0.0	0	2015	Albany

Figure 3-13. *Output dataframe*

Step 5-2. Get the data in the required format.

```
data = df[['Date', 'Total Volume', '4046', '4225', '4770',
'Small Bags', 'type']]
data.rename(columns={'Date':'ds'},inplace=True)
data['y'] = df.iloc[:,1]
```

Step 5-3. Do a train-test split.

```
train_df = data[:18000]
test_df = data[18000:]
```

Step 5-4. Initialize the Prophet model and add a regressor.

```
prophet_model_regressor = Prophet()
prophet_model_regressor.add_regressor('type')
prophet_model_regressor.add_regressor('Total Volume')
```

```
prophet_model_regressor.add_regressor('4046')
prophet_model_regressor.add_regressor('4225')
prophet_model_regressor.add_regressor('4770')
prophet_model_regressor.add_regressor('Small Bags')
```

Step 5-5. Fit the data.

```
prophet_model_regressor.fit(train_df)
future_df = prophet_model_regressor.make_future_
dataframe(periods=249)
```

Step 5-6. Forecast the data in the test.

```
forecast_df = prophet_model_regressor.predict(test_df)
prophet_model_regressor.plot(forecast_df)
```

Figure 3-14 shows the forecast output.

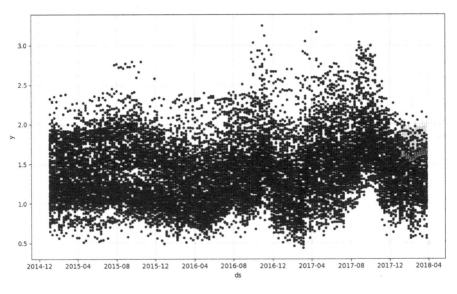

Figure 3-14. *Forecast output*

Figure 3-14 shows that the blue area is the predicted/forecasted data.

Recipe 3-6. Time Series Forecasting Using Auto-ARIMA

Problem

You want to load time series data and forecast using Auto-ARIMA.

Solution

It can be easily achieved using the built-in method defined in the stats model.

How It Works

The following steps read the data and forecast using Auto-ARIMA.

Step 6-1. Import the required libraries.

```
#import all the required libraries
import pandas as pd
from pmdarima.arima import auto_arima
from pmdarima.arima import ADFTest
from matplotlib import pyplot as plt
from sklearn.metrics import r2_score
```

Step 6-2. Read the data.

Download the data from the Git link.

The following reads the data.

```
#read data
auto_arima_data = pd.read_csv('auto_arima_data.txt')
auto_arima_data.head()
```

Figure 3-15 shows the output dataframe.

	Month	Champagne sales
0	1964-01	2815
1	1964-02	2672
2	1964-03	2755
3	1964-04	2721
4	1964-05	2946

Figure 3-15. *Output dataframe*

After reading the data, let's ensure there are no nulls present in the data. And check the datatype of each column.

Step 6-3. Preprocess the data.

```
#check missing values
auto_arima_data.isnull().sum()
```

The output is as follows.

```
Month            0
Champagne sales  0
dtype: int64
```

There are no nulls present. Let's check the datatype of each column.

```
#check datatype
auto_arima_data.info()
```

The output is as follows.

```
<class 'pandas.core.frame.DataFrame'>
RangeIndex: 105 entries, 0 to 104
Data columns (total 2 columns):
 #   Column            Non-Null Count  Dtype
---  ------            --------------  -----
 0   Month             105 non-null    object
 1   Champagne sales   105 non-null    int64
dtypes: int64(1), object(1)
memory usage: 1.8+ KB
```

Now let's change the datatype of the 'Month' variable from string to datetime. Also, set the dataframe index to this variable using the set_index method.

```
#convert object to datatime and set index
auto_arima_data['Month'] = pd.to_datetime(auto_arima_data['Month'])
auto_arima_data.set_index('Month',inplace=True)
auto_arima_data.head()
```

Figure 3-16 shows the output dataframe.

	Champagne sales
Month	
1964-01-01	2815
1964-02-01	2672
1964-03-01	2755
1964-04-01	2721
1964-05-01	2946

Figure 3-16. *Output dataframe*

Step 6-4. Analyze the data pattern.

You can simply plot the line chart using the .plot() method to analyze the basic time series components.

```
#line plot to understand the pattern
auto_arima_data.plot()
```

Figure 3-17 shows the time series plot output.

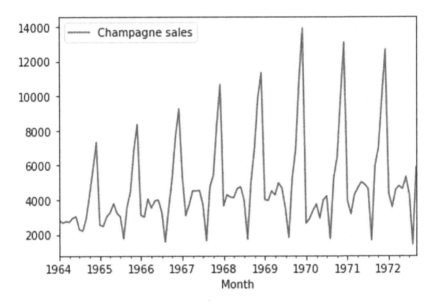

Figure 3-17. *Output*

The chart shows that there is a seasonality associated with the target column. There is a spike in sales every year.

Step 6-5. Test for stationarity.

A stationarity check is an important step in a time series use case if you use any statistical model approach. Let's use the Augmented Dickey-Fuller (ADF) test to check the data's stationarity.

```
#Stationarity check
stationary_test = ADFTest(alpha= 0.05)
stationary_test.should_diff(auto_arima_data)
```

The output is as follows.

```
(0.01, False)
```

The output shows that the data is non-stationary. So let's make the data stationary while building the Auto-ARIMA model. The integrated (I) concept, denoted by the 'd' value, is used.

Step 6-6. Split the dataset to train and test.

Let's split the dataset into two parts: train and test. Build or train the model using the training set, and forecast the data using the testing set.

```
#train test split and plot
train_data = auto_arima_data[:85]
test_data = auto_arima_data[-20:]
plt.plot(train_data)
plt.plot(test_data)
```

Figure 3-18 shows the train-test split output.

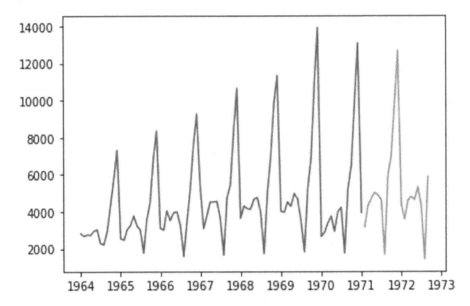

Figure 3-18. *Train-test split output*

In Figure 3-18, the blue line represents the training set, and the orange line represents the test set.

Step 6-7. Build the Auto-ARIMA model.

In the Auto-ARIMA model, the lowercase p, d, and q values indicate the non-seasonal components. The uppercase P, D, and Q values indicate seasonal components. It works similarly to hypertuning techniques to find the optimal value of p, d, and q with different combinations. The final values are determined with the lower AIC and BIC parameters considered.

Let's try p, d, and q values ranging from 0 to 5 to get the best values from the model.

```
#model building with parameters

auto_arima_model = auto_arima(train_data, start_p = 0, d=1,
start_q = 0, max_p = 5, max_d = 5,
```

max_q= 5, start_P = 0, D=1,
start_Q = 0, max_P = 5, max_D = 5,
max_Q= 5, m=12, seasonal = True,
error_action = 'warn', trace = True, supress_warnings= True,
stepwise = True, random_state =20,
n_fits = 50)

Figure 3-19 shows the Auto-ARIMA output.

```
Performing stepwise search to minimize aic
 ARIMA(0,1,0)(0,1,0)[12]             : AIC=1203.853, Time=0.03 sec
 ARIMA(1,1,0)(1,1,0)[12]             : AIC=1192.025, Time=0.11 sec
 ARIMA(0,1,1)(0,1,1)[12]             : AIC=1176.246, Time=0.26 sec
 ARIMA(0,1,1)(0,1,0)[12]             : AIC=1174.731, Time=0.10 sec
 ARIMA(0,1,1)(1,1,0)[12]             : AIC=1176.034, Time=0.24 sec
 ARIMA(0,1,1)(1,1,1)[12]             : AIC=1176.700, Time=0.43 sec
 ARIMA(1,1,1)(0,1,0)[12]             : AIC=1175.054, Time=0.17 sec
 ARIMA(0,1,2)(0,1,0)[12]             : AIC=1174.769, Time=0.15 sec
 ARIMA(1,1,0)(0,1,0)[12]             : AIC=1194.721, Time=0.05 sec
 ARIMA(1,1,2)(0,1,0)[12]             : AIC=1174.564, Time=0.30 sec
 ARIMA(1,1,2)(1,1,0)[12]             : AIC=inf, Time=0.49 sec
 ARIMA(1,1,2)(0,1,1)[12]             : AIC=inf, Time=0.42 sec
 ARIMA(1,1,2)(1,1,1)[12]             : AIC=1176.859, Time=0.84 sec
 ARIMA(2,1,2)(0,1,0)[12]             : AIC=1176.127, Time=0.31 sec
 ARIMA(1,1,3)(0,1,0)[12]             : AIC=1176.124, Time=0.41 sec
 ARIMA(0,1,3)(0,1,0)[12]             : AIC=1176.458, Time=0.19 sec
 ARIMA(2,1,1)(0,1,0)[12]             : AIC=1176.656, Time=0.18 sec
 ARIMA(2,1,3)(0,1,0)[12]             : AIC=1180.597, Time=0.47 sec
 ARIMA(1,1,2)(0,1,0)[12] intercept   : AIC=inf, Time=0.23 sec

 Best model:   ARIMA(1,1,2)(0,1,0)[12]
 Total fit time: 5.376 seconds
```

Figure 3-19. *Auto-ARIMA output*

The following is the summary of the model.

```
#model summary
auto_arima_model.summary()
```

Figure 3-20 shows the Auto-ARIMA model summary.

SARIMAX Results

Dep. Variable:		y		**No. Observations:**		85
Model:	SARIMAX(1, 1, 2)x(0, 1, [], 12)			**Log Likelihood**		-583.282
Date:		Mon, 19 Sep 2022		**AIC**		1174.564
Time:		22:55:59		**BIC**		1183.670
Sample:		01-01-1964		**HQIC**		1178.189
		- 01-01-1971				
Covariance Type:		opg				

	coef	std err	z	P>|z|	[0.025	0.975]
ar.L1	-0.8412	0.152	-5.543	0.000	-1.139	-0.544
ma.L1	0.0513	0.167	0.308	0.758	-0.275	0.378
ma.L2	-0.8673	0.086	-10.134	0.000	-1.035	-0.700
sigma2	5.862e+05	7.03e+04	8.342	0.000	4.48e+05	7.24e+05

Ljung-Box (L1) (Q):	0.05	**Jarque-Bera (JB):**	8.55
Prob(Q):	0.83	**Prob(JB):**	0.01
Heteroskedasticity (H):	2.61	**Skew:**	-0.10
Prob(H) (two-sided):	0.02	**Kurtosis:**	4.68

Figure 3-20. *Auto-ARIMA model summary*

Step 6-8. Forecast using the test data.

Let's forecast using the test set.

```
#forecasting on test set
pred = pd.DataFrame(auto_arima_model.predict(n_periods = 20),
index = test_data.index)
pred.columns= ['pred_sales']

#plot
```

```
plt.figure(figsize=(8,5))
plt.plot(train_data, label = "Training data")
plt.plot(test_data, label = "Test data")
plt.plot(pred, label = "Predicted data")
plt.legend()
plt.show()
```

Figure 3-21 shows the forecast output.

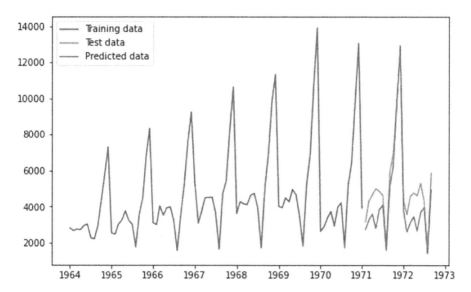

Figure 3-21. *Forecast output*

Step 6-9. Evaluate the model.

Let's evaluate the model using test set predictions.

```
#Evaluating using r square score
test_data['prediction'] = pred
r2_score(test_data['Champagne sales'],test_data['prediction'])
```

The output is as follows.

```
0.811477383636726
```

Recipe 3-7. Multivariate Time Series Forecasting Using the VAR Model

Problem

You want to load the time series data and forecast using multiple features.

Solution

It can be easily achieved using the Vector Auto Regressive (VAR) model defined in the stats model.

How It Works

The following steps read the data and forecast using the VAR model.

Step 7-1. Import the required libraries.

```
#import all the required libraries
import pandas as pd
from math import sqrt
from sklearn.metrics import mean_squared_error
from statsmodels.tsa.vector_ar.vecm import coint_johansen
from statsmodels.tsa.vector_ar.var_model import VAR
```

Step 7-2. Read the data.

Download the data from the Git link.

The following reads the data.

```
#read data
var_data = pd.read_excel('AirQualityUCI.xlsx', parse_
dates=[['Date', 'Time']])

var_data.head()
```

While reading the data, let's parse the datetime columns.

Figure 3-22 shows the output dataframe.

Date_Time	CO(GT)	PT08.S1(CO)	NMHC(GT)	C6H6(GT)	PT08.S2(NMHC)	NOx(GT)	PT08.S3(NOx)	NO2(GT)	PT08.S4(NO2)	PT08.S5(O3)	T	RI
0	2004-03-10 18:00:00	2.6	1360.00	150	11.881723	1045.50	166.0	1056.25	113.0	1692.00	1267.50	13.60	48.87500
1	2004-03-10 19:00:00	2.0	1292.25	112	9.397165	954.75	103.0	1173.75	92.0	1558.75	972.25	13.30	47.70000
2	2004-03-10 20:00:00	2.2	1402.00	88	8.997817	939.25	131.0	1140.00	114.0	1554.50	1074.00	11.90	53.97500
3	2004-03-10 21:00:00	2.2	1375.50	80	9.228796	948.25	172.0	1092.00	122.0	1583.75	1203.25	11.00	60.00000
4	2004-03-10 22:00:00	1.6	1272.25	51	6.518224	835.50	131.0	1205.00	116.0	1490.00	1110.00	11.15	59.57500

Figure 3-22. *Output dataframe*

After reading the data, let's ensure there are no nulls.

Step 7-3. Preprocess the data.

```
#check missing values

var_data.isnull().sum()
```

The output is as follows.

```
Date_Time        0
CO(GT)           0
PT08.S1(CO)      0
NMHC(GT)         0
C6H6(GT)         0
PT08.S2(NMHC)    0
NOx(GT)          0
PT08.S3(NOx)     0
NO2(GT)          0
PT08.S4(NO2)     0
PT08.S5(O3)      0
T                0
```

```
RH                   0
AH                   0
dtype: int64
```

Now let's change the datatype of the Date_Time variable from string to datetime. Also, set the index of the dataframe to this variable using the set_index method.

```
var_data['Date_Time'] = pd.to_datetime(var_data.Date_Time ,
format = '%d/%m/%Y %H.%M.%S')
var_data1 = var_data.drop(['Date_Time'], axis=1)
var_data1.index = var_data.Date_Time
var_data1.head()
```

Figure 3-23 shows the output dataframe.

Date_Time	CO(GT)	PT08.S1(CO)	NMHC(GT)	C6H6(GT)	PT08.S2(NMHC)	NOx(GT)	PT08.S3(NOx)	NO2(GT)	PT08.S4(NO2)	PT08.S5(O3)	T	RH
2004-03-10 18:00:00	2.6	1360.00	150	11.881723	1045.50	166.0	1056.25	113.0	1692.00	1267.50	13.60	48.875001
2004-03-10 19:00:00	2.0	1292.25	112	9.397165	954.75	103.0	1173.75	92.0	1558.75	972.25	13.30	47.700000
2004-03-10 20:00:00	2.2	1402.00	88	8.997817	939.25	131.0	1140.00	114.0	1554.50	1074.00	11.90	53.975000
2004-03-10 21:00:00	2.2	1375.50	80	9.228796	948.25	172.0	1092.00	122.0	1583.75	1203.25	11.00	60.000000
2004-03-10 22:00:00	1.6	1272.25	51	6.518224	835.50	131.0	1205.00	116.0	1490.00	1110.00	11.15	59.575001

Figure 3-23. *Output dataframe*

Currently, there are no missing values present in the dataset. To be on the safe side, let's have code in place in case you encounter any nulls in your dataset.

```
#missing value treatment
cols = var_data1.columns
for j in cols:
```

```
for i in range(0,len(var_data1)):
    if var_data1[j][i] == -200:
        var_data1[j][i] = var_data1[j][i-1]
```

Step 7-4. Check the stationarity.

Let's use the ADF test to check the stationarity of the data. This test works on a maximum of 12 variables, so let's randomly drop one since there are 13.

```
#checking stationarity
from statsmodels.tsa.vector_ar.vecm import coint_johansen
#since the test works for only 12 variables, I have
randomly dropped
#in the next iteration, I would drop another and check the
eigenvalues
test = var_data1.drop([ 'CO(GT)'], axis=1)
coint_johansen(test,-1,1).eig
```

The output is as follows.

```
array([1.75628733e-01, 1.52399674e-01, 1.15090998e-01,
1.04309966e-01,
       9.29562919e-02, 6.90255307e-02, 5.76654697e-02,
3.43596700e-02,
       3.06350634e-02, 1.18801270e-02, 2.46819409e-03,
7.09473977e-05])
```

Step 7-5. Split the dataset into train-test.

Let's split the dataset.

```
#creating the train and validation set
train_data = var_data1[:int(0.8*(len(var_data1)))]
valid_data = var_data1[int(0.8*(len(var_data1))):]
```

Step 7-6. Build the VAR model and forecast on the test set.

```
##fit the model
from statsmodels.tsa.vector_ar.var_model import VAR

var_model = VAR(endog=train_data)
var_model_fit = var_model.fit()

# make prediction on validation
pred = var_model_fit.forecast(var_model_fit.endog,
steps=len(valid_data))

pred
```

Figure 3-24 shows the prediction output.

```
array([[8.88161065e-01, 8.41803964e+02, 2.71644320e+02, ...,
        1.05743863e+01, 3.48713152e+01, 4.37277520e-01],
       [9.92424381e-01, 8.66262441e+02, 2.69327633e+02, ...,
        9.85432359e+00, 3.74025472e+01, 4.42645045e-01],
       [1.10490663e+00, 8.90900736e+02, 2.67743663e+02, ...,
        9.24271941e+00, 3.96241504e+01, 4.47398014e-01],
       ...,
       [2.13383727e+00, 1.10685527e+03, 2.69411125e+02, ...,
        2.01829088e+01, 4.88992477e+01, 1.11131218e+00],
       [2.13383726e+00, 1.10685527e+03, 2.69411126e+02, ...,
        2.01829090e+01, 4.88992475e+01, 1.11131220e+00],
       [2.13383724e+00, 1.10685527e+03, 2.69411126e+02, ...,
        2.01829093e+01, 4.88992474e+01, 1.11131221e+00]])
```

Figure 3-24. *Prediction output*

The predictions are in the form of an array, where each list represents the predictions of the row. Let's transform this into a more presentable format.

```
##converting predictions to dataframe
pred1 = pd.DataFrame(index=range(0,len(pred)),columns=[cols])
for j in range(0,13):
```

```
for i in range(0, len(pred1)):
    pred1.iloc[i][j] = pred[i][j]
pred1
```

Figure 3-25 shows the output dataframe.

	CO(GT)	PT08.S1(CO)	NMHC(GT)	C6H6(GT)	PT08.S2(NMHC)	NOx(GT)	PT08.S3(NOx)	NO2(GT)	PT08.S4(NO2)	PT08.S5(O3)	T	RH
0	0.888161	841.803964	271.64432	1.982632	595.965721	137.03066	1119.615156	87.815339	829.84466	546.026663	10.574386	34.871
1	0.992424	866.262441	269.327633	2.332532	619.914837	158.117747	1098.381882	90.537193	856.405973	598.407982	9.854324	37.402
2	1.104907	890.900736	267.743663	2.832077	645.634216	177.729196	1078.45843	93.115608	885.665916	648.974793	9.242719	39.62
3	1.219987	914.919521	266.564032	3.407752	671.662218	195.906115	1059.928335	95.57167	915.647307	697.359008	8.722839	41.579
4	1.333743	937.819088	265.921087	4.011249	697.061083	212.682873	1042.817739	97.910723	945.09167	743.269637	8.281647	43.304
...	...	...	...	...	...	...	...	...	...	...	...	...
1867	2.133837	1106.855274	269.411125	10.807054	966.524358	228.625867	843.010179	101.410243	1534.884856	1039.936425	20.182908	48.899
1868	2.133837	1106.855272	269.411125	10.807054	966.524357	228.625864	843.010179	101.410243	1534.884861	1039.93642	20.182909	48.899
1869	2.133837	1106.85527	269.411125	10.807054	966.524357	228.625861	843.010179	101.410242	1534.884866	1039.936415	20.182909	48.899
1870	2.133837	1106.855268	269.411126	10.807054	966.524356	228.625857	843.010179	101.410242	1534.884871	1039.936411	20.182909	48.899
1871	2.133837	1106.855266	269.411126	10.807054	966.524355	228.625854	843.010179	101.410241	1534.884876	1039.936406	20.182909	48.899

1872 rows × 13 columns

Figure 3-25. *Output dataframe*

Step 7-7. Evaluate the model.

Let's get the RMSE (root-mean-square error) evaluation metrics for each variable.

```
##check rmse
for i in cols:
    print('rmse value for', i, 'is : ', sqrt(mean_squared_
error(pred1[i], valid_data[i])))
```

The output is as follows.

```
rmse value for CO(GT) is :  1.4086965424457896
rmse value for PT08.S1(CO) is :  205.91037633777376
rmse value for NMHC(GT) is :  6.670741427642936
rmse value for C6H6(GT) is :  7.130304477786223
rmse value for PT08.S2(NMHC) is :  277.8562837309765
rmse value for NOx(GT) is :  214.7579379769933
```

```
rmse value for PT08.S3(NOx) is :  244.9612992895686
rmse value for NO2(GT) is :  66.65226538131333
rmse value for PT08.S4(NO2) is :  490.052866528993
rmse value for PT08.S5(O3) is :  446.50499189012726
rmse value for T is :  10.722429361274823
rmse value for RH is :  17.114848634832306
rmse value for AH is :  0.5216105887695865
```

CHAPTER 4

Machine Learning Regression–based Forecasting

The previous chapters explained how to forecast future values using time series algorithms. Again, in time series modeling, there are two types of time series: univariate and multivariate. For more information, please refer to Chapters 2 and 3.

This chapter aims to build classical machine learning (ML) regression algorithms for time series forecasting. Machine learning–based forecasting is powerful because forecasting takes other factors/features to forecast the values.

The chapter focuses on building regressor models for forecasting.

Recipe 4-1. Formulating Regression Modeling for Time Series Forecasting

Recipe 4-2. Implementation of the XGBoost Model

Recipe 4-3. Implementation of the LightGBM Model

Recipe 4-4. Implementation of the Random Forest Model

Recipe 4-5. Selecting the Best Model

© Akshay R Kulkarni, Adarsha Shivananda, Anoosh Kulkarni, V Adithya Krishnan 2023
A. R. Kulkarni et al., *Time Series Algorithms Recipes*,
https://doi.org/10.1007/978-1-4842-8978-5_4

Recipe 4-1. Formulating Regression Modeling for Time Series Forecasting

Problem

You want to formulate regression models for time series forecasting.

Solution

The following are the basic steps for machine learning regression–based forecasting.

1. Data collection

2. Data cleaning and preprocessing

3. Feature selection

4. Train–test–validation split

5. Model building

6. Evaluation

7. Prediction

Figure 4-1 shows the steps to build ML regression–based forecasting.

Figure 4-1. *Output*

How It Works

The following steps build regressor models.

Step 1-1. Install and import the required libraries.

Let's import all the required libraries.

```
#import libraries
import pandas as pd
import numpy as np
import glob
import time
from sklearn.feature_selection import SelectKBest
from sklearn.feature_selection import f_regression
import xgboost as xgb
from lightgbm import LGBMRegressor
from sklearn.ensemble import RandomForestRegressor
from sklearn.metrics import mean_squared_error
from plotly.offline import init_notebook_mode, iplot
import plotly.graph_objs as go
from plotly import tools
init_notebook_mode(connected=True)
from sklearn.metrics import mean_squared_error
```

Step 1-2. Collect the data.

This chapter uses electricity consumption data. Please find the dataset on GitHub.

The following code reads the data.

```
df = pd.read_csv('train_6BJx641.csv')
df.head()
```

Figure 4-2 shows the first five rows of the data.

	ID	datetime	temperature	var1	pressure	windspeed	var2	electricity_consumption
0	0	2013-07-01 00:00:00	-11.4	-17.1	1003.0	571.910	A	216.0
1	1	2013-07-01 01:00:00	-12.1	-19.3	996.0	575.040	A	210.0
2	2	2013-07-01 02:00:00	-12.9	-20.0	1000.0	578.435	A	225.0
3	3	2013-07-01 03:00:00	-11.4	-17.1	995.0	582.580	A	216.0
4	4	2013-07-01 04:00:00	-11.4	-19.3	1005.0	586.600	A	222.0

Figure 4-2. *Output*

Step 1-3. Preprocess the data and create features (feature engineering).

Before building any model, checking the data quality and preprocessing as per the model requirement is a must.

So, let's see what cleaning and preprocessing are required.

The dataset consists of an ID column that cannot be fed into the model, so let's drop the ID column first.

```
del df['ID']
```

Let's check the missing values.

```
df.isnull().sum()
```

Figure 4-3 shows the output of the nulls.

The output is as follows.

```
datetime                  0
temperature               0
var1                      0
pressure                  0
windspeed                 0
var2                      0
electricity_consumption   0
dtype: int64
```

Figure 4-3. *Output*

There are no missing values present in any of the columns, so treating them is not required.

Now, let's check the datatype of each column.

```
df.info()
```

Figure 4-4 shows the output of each column datatype.

```
<class 'pandas.core.frame.DataFrame'>
RangeIndex: 26496 entries, 0 to 26495
Data columns (total 7 columns):
 #   Column                   Non-Null Count   Dtype
---  ------                   --------------   -----
 0   datetime                 26496 non-null   object
 1   temperature              26496 non-null   float64
 2   var1                     26496 non-null   float64
 3   pressure                 26496 non-null   float64
 4   windspeed                26496 non-null   float64
 5   var2                     26496 non-null   object
 6   electricity_consumption  26496 non-null   float64
dtypes: float64(5), object(2)
memory usage: 1.4+ MB
```

Figure 4-4. *Output*

Let's convert the date column into the pandas datetime format and create features (known as *feature engineering*).

```
#Creating datetime features to use in the model to capture
seasonality
df['time'] = pd.to_datetime(df['datetime'])
df['year'] = df.time.dt.year
df['month'] = df.time.dt.month
df['day'] = df.time.dt.day
df['hour'] = df.time.dt.hour
df.drop('time', axis=1, inplace=True)
df.head()
```

Figure 4-5 shows the output of the first five rows after creating datetime features.

	datetime	temperature	var1	pressure	windspeed	var2	electricity_consumption	year	month	day	hour
0	2013-07-01 00:00:00	-11.4	-17.1	1003.0	571.910	A	216.0	2013	7	1	0
1	2013-07-01 01:00:00	-12.1	-19.3	996.0	575.040	A	210.0	2013	7	1	1
2	2013-07-01 02:00:00	-12.9	-20.0	1000.0	578.435	A	225.0	2013	7	1	2
3	2013-07-01 03:00:00	-11.4	-17.1	995.0	582.580	A	216.0	2013	7	1	3
4	2013-07-01 04:00:00	-11.4	-19.3	1005.0	586.600	A	222.0	2013	7	1	4

Figure 4-5. *Output*

Let's sort the dataframe using the datetime column and then drop it because it is not considered in model building.

```
#sorting
df=df.sort_values(by='datetime')

#Deleting the column
del df['datetime']
```

Figure 4-6 shows the output of the first five rows after sorting and deleting the datetime column.

	temperature	var1	pressure	windspeed	var2	electricity_consumption	year	month	day	hour
0	-11.4	-17.1	1003.0	571.910	A	216.0	2013	7	1	0
1	-12.1	-19.3	996.0	575.040	A	210.0	2013	7	1	1
2	-12.9	-20.0	1000.0	578.435	A	225.0	2013	7	1	2
3	-11.4	-17.1	995.0	582.580	A	216.0	2013	7	1	3
4	-11.4	-19.3	1005.0	586.600	A	222.0	2013	7	1	4

Figure 4-6. *Output*

The year, month, day, and hour columns are created using the datetime column and considering them to build the model to capture trend and seasonality. Likewise, you can add lag features of the target variable as features.

Piece of code to create lag features: df['lag1']=df['col name'].shift(1)

The dataset consists of one categorical feature so let's encode using the one-hot method.

```
#convering all categorical columns to numerical.
df1=pd.get_dummies(df)
```

Figure 4-7 shows the output of the first five rows after encoding.

	temperature	var1	pressure	windspeed	electricity_consumption	year	month	day	hour	var2_A	var2_B	var2_C
0	-11.4	-17.1	1003.0	571.910	216.0	2013	7	1	0	1	0	0
1	-12.1	-19.3	996.0	575.040	210.0	2013	7	1	1	1	0	0
2	-12.9	-20.0	1000.0	578.435	225.0	2013	7	1	2	1	0	0
3	-11.4	-17.1	995.0	582.580	216.0	2013	7	1	3	1	0	0
4	-11.4	-19.3	1005.0	586.600	222.0	2013	7	1	4	1	0	0

Figure 4-7. *Output*

The get_dummies method is used here. It is defined in pandas to encode all categorical columns into numerical ones. There is only one column in this dataset.

The required cleaning and preprocessing are done. Next, let's focus on feature selection.

Step 1-4. Select the features.

To build supervised ML models, you need to consider those features that are significant to the target feature.

To get the significant features to build the model, you can perform advanced statistical tests like Anova, chi-squared, and correlation, or there is a built-in function defined in sklearn called SelectKBest, which helps you select features by providing scores for each column.

Before using SelectKBest, let's create separate objects for all independent and target features.

```
#creating target and features objects
x = df1.drop(columns=['electricity_consumption'])
y = df1.iloc[:,4]
```

Now, let's fit SelectKBest.

```
#implementing selectKbest
st=time.time()
bestfeatures = SelectKBest(score_func=f_regression)
fit = bestfeatures.fit(x,y)
et=time.time()-st
print(et)
dfscores = pd.DataFrame(fit.scores_)
dfcolumns = pd.DataFrame(x.columns)
featureScores = pd.concat([dfcolumns,dfscores],axis=1)
featureScores.columns = ['Featuress','Score']
best_features=featureScores.nlargest(5,'Score')
best_features
```

Figure 4-8 shows the output of the top five features.

	Featuress	Score
3	windspeed	1603.378268
1	var1	483.788043
0	temperature	369.330583
6	day	256.724699
9	var2_B	74.551514

Figure 4-8. *Output*

The n value is 5, so five features with a score are shown.

Step 1-5. Work on the train-test and validation split.

The feature selection part is done. So now, let's split the dataset for model building. You can use the built-in function to split the data. But since this is a forecasting problem, you don't want it to split randomly. The latest date records should go into the validation set. Beginning records should go into the training set. Middle records into the test set.

Let's use the basic pandas method to split the data.

```
# train-test-validation split
test=df1.tail(7940)
#test set
test1=test.head(7440)
#training set
train=df1.head(18556)
#validation set
pred=test.tail(500)
```

Since the data is prepared and in the correct format, let's start with the modeling process.

Let's try multiple models and then choose the one with the best performance.

Note that all features are considered when building the models, irrespective of the feature selection output.

Before fitting the model, again, create separate objects for all independent and target features.

```
#creating target and features objects
#for training
y_train=train.iloc[:,4]
X_train=train.drop(columns=['electricity_consumption'])

#for test
y_test=test1.iloc[:,4]
X_test=test1.drop(columns=['electricity_consumption'])

#for validation
y_pred=pred.iloc[:,4]
X_pred=pred.drop(columns=['electricity_consumption'])
```

Recipe 4-2. Implementing the XGBoost Model

Problem

You want to use the XGBoost model.

Solution

The simplest way to build is to use the sklearn library.

How It Works

Let's follow the steps.

Step 2-1. Build the XGBoost model.

The preprocessed train data is ready to build the model (from Recipe 4-1). Let's build the XGBoost model.

```
# Xgboost model

xg_reg = xgb.XGBRegressor(objective ='reg:linear', colsample_
bytree = 0.3, learning_rate = 0.1,
                max_depth = 100, alpha = 10, n_
estimators = 140)
xg_reg.fit(X_train,y_train)
```

Step 2-2. Evaluate the XGBoost model in the test set.

```
# Evaluating the model on test data
predictions = xg_reg.predict(X_test)
errors = abs(predictions - y_test)
mape = 100 * np.mean(errors / y_test)
```

```
mse=mean_squared_error(y_test,predictions)
RMSE=np.sqrt(mse)
print("XGBOOST model")
print("mape value for test set",mape)
print("mse value for test set",mse)
print("RMSE value for test set",RMSE)
```

The output is as follows.

```
XGBOOST model
mape value for test set 20.395743624282215
mse value for test set 9949.946409810507
RMSE value for test set 99.74941809259093
```

Step 2-3. Evaluate the XGBoost model in the validation set.

```
# Evaluating the model on test data

predictions = xg_reg.predict(X_pred)
errors = abs(predictions - y_pred)
mape = 100 * np.mean(errors / y_pred)
mse=mean_squared_error(y_pred,predictions)
RMSE=np.sqrt(mse)
print("XGBOOST model")
print("mape value for validation set",mape)
print("mse value for validation set",mse)
print("RMSE value for validation set",RMSE)
```

The output is as follows.

```
XGBOOST model
mape value for validation set 18.31413324392246
mse value for validation set 6692.219349248934
RMSE value for validation set 81.80598602332799
```

Recipe 4-3. Implementing the LightGBM Model

Problem

You want to use the LightGBM model.

Solution

The simplest way to build is using the sklearn library.

How It Works

Let's follow the steps.

Step 3-1. Build the LightGBM model.

The same preprocessed train data (from Recipe 4-1) is used to build the LightGBM model.

```
# LightGBM model

lgb_reg = LGBMRegressor(n_estimators=100, random_state=42)
lgb_reg.fit(X_train, y_train)
```

Step 3-2. Evaluate the LightGBM model in the test set.

```
# Evaluating the model on test data
predictions = lgb_reg.predict(X_test)
errors = abs(predictions - y_test)
mape = 100 * np.mean(errors / y_test)
mse=mean_squared_error(y_test,predictions)
RMSE=np.sqrt(mse)
```

```
print("LIGHTGBM model")
print("mape value for test set",mape)
print("mse value for test set",mse)
print("RMSE value for test set",RMSE)
```

The output is as follows.

```
LIGHTGBM model
mape value for test set 17.8086387209238
mse value for test set 7448.058075387331
RMSE value for test set 86.30213250776212
```

Step 3-3. Evaluate the LightGBM model in the validation set.

```
# Evaluating the model on test data
predictions = lgb_reg.predict(X_pred)
errors = abs(predictions - y_pred)
mape = 100 * np.mean(errors / y_pred)
mse=mean_squared_error(y_pred,predictions)
RMSE=np.sqrt(mse)
print("LIGHTGBM model")
print("mape value for validation set",mape)
print("mse value for validation set",mse)
print("RMSE value for validation set",RMSE)
```

The output is as follows.

```
LIGHTGBM model
mape value for validation set 14.524462046915062
mse value for validation set 4610.576774339071
RMSE value for validation set 67.90122807681074
```

Recipe 4-4. Implementing the Random Forest Model

Problem

You want to use the random forest model.

Solution

The simplest way to build is using the sklearn library.

How It Works

Let's follow the steps.

Step 4-1. Build a random forest model.

The same preprocessed train data (from Recipe 4-1) is used to build the random forest model.

```
# Random Forest model

regr = RandomForestRegressor(n_estimators=100, random_state=42)
regr.fit(X_train, y_train)
```

Step 4-2. Evaluate the LightGBM model in the test set.

```
# Evaluating the model on test data
predictions = regr.predict(X_test)
errors = abs(predictions - y_test)
mape = 100 * np.mean(errors / y_test)
mse=mean_squared_error(y_test,predictions)
RMSE=np.sqrt(mse)
```

```
print("RANDOM FOREST model")
print("mape value for test set",mape)
print("mse value for test set",mse)
print("RMSE value for test set",RMSE)
```

The output is as follows.

```
RANDOM FOREST model
mape value for test set 18.341864621229462
mse value for test set 7642.701889959678
RMSE value for test set 87.42254794936875
```

Step 4-3. Evaluate the LightGBM model in the validation set.

```
# Evaluating the model on test data
predictions = regr.predict(X_pred)
errors = abs(predictions - y_pred)
mape = 100 * np.mean(errors / y_pred)
mse=mean_squared_error(y_pred,predictions)
RMSE=np.sqrt(mse)
print("RANDOM FOREST model")
print("mape value for validation set",mape)
print("mse value for validation set",mse)
print("RMSE value for validation set",RMSE)
```

The output is as follows.

```
RANDOM FOREST model
mape value for validation set 16.41982254170068
mse value for validation set 5138.454886199999
RMSE value for validation set 71.68301672083841
```

Recipe 4-5. Selecting the Best Model

Problem

You want to select the best-performing model out of all the ones implemented.

Solution

The following are some of the evaluation metrics used.

- MAE (mean absolute error) is calculated as the mean of the absolute error term, where the error is the difference between the actuals and the predictions.

- MSE (mean squared error) is the mean of the squared error term, where the error is the difference between the actuals and the predictions.

- RMSE (root mean square error) is the square root of the mean square error.

- MAPE (mean absolute percentage error) is the mean of the absolute percentage errors, where the percentage error is the ratio between the error and actuals.

- Accuracy is calculated by subtracting one with mean absolute percentage error in time series and regression.

All three regressor models are trained. Instead of looking at the evaluation metrics of each model on every model training, let's use a piece of code that provides the best model for predictions and plots the predictions against actuals.

How It Works

The following steps select the best-performing model.

Step 5-1. Evaluate the method.

Let's write a function to evaluate these models.

```
#function to evaluate the model.
def evaluate(model, test_features, test_labels):
    predictions = model.predict(test_features)
    errors = abs(predictions - test_labels)
    mape = 100 * np.mean(errors / test_labels)
    accuracy = 100 - mape
    mse=mean_squared_error(test_labels,predictions)
    RMSE=np.sqrt(mse)
    print('Model Performance')
    print('Average Error: {:0.4f} degrees.'.format(np.
mean(errors)))
    print('Accuracy = {:0.2f}%.'.format(accuracy))
    print('RMSE = {:0.2f}'.format(RMSE))
    return accuracy,predictions,RMSE
```

Step 5-2. Compare performance in the test set.

Let's call evaluate function to get the best performance model, which returns the best model name, model object, and predictions.

Let's check the best performance model for the test set.

```
models=[xg_reg,lgb_reg,regr]
model_name=['XGBoost','LightGBM','RandomForest']
model_RMSE=[]
model_predictions=[]
for item in models:
```

```
    base_accuracy,predictions,RMSE=evaluate(item,X_test,y_test)
    model_RMSE.append(RMSE)
    model_predictions.append(predictions)
r=model_RMSE.index(min(model_RMSE))
best_model_predictions=model_predictions[r]
best_model_name=model_name[r]
best_model=models[r]
```

The output is as follows.

```
Model Performance
Average Error: 67.7595 degrees.
Accuracy = 79.60%.
RMSE = 99.75
Model Performance
Average Error: 58.4586 degrees.
Accuracy = 82.19%.
RMSE = 86.30
Model Performance
Average Error: 59.3871 degrees.
Accuracy = 81.66%.
RMSE = 87.42
```

```
print('Best Model:')
print(best_model_name)
print('Model Object:')
print(best_model)
print('Predictions:')
print(best_model_predictions)
```

Figure 4-9 shows the output of the best model, object, and predictions for the test set.

The output is as follows.

```
Best Model:
LightGBM
Model Object:
LGBMRegressor(random_state=42)
Predictions:
[216.59204245 232.93339549 224.6671183  ... 198.45734021 192.5073198
  191.79087945]
```

Figure 4-9. *Output*

Among the three regressor models, LightGBM performs the best in both test sets. It has the lowest RMSE (86).

Let's use the LightGBM model predictions.

Step 5-3. Plot the LightGBM model prediction against the actuals in the test set.

Let's plot predictions vs. actuals charts using Plotly for the test set

```
#Plot timeseries
y_test=pd.DataFrame(y_test)

y_test['predictions']=best_model_predictions

X_test['datetime']=pd.to_datetime(X_test[['year','month','day',
'hour']])

y_test['datetime']=X_test['datetime']

y_test=y_test.sort_values(by='datetime')

trace0 = go.Scatter(x=y_test['datetime'].astype(str),
y=y_test['electricity_consumption'].values, opacity = 0.8,
name='actual_value')
trace1 = go.Scatter(x=y_test['datetime'].astype(str), y=y_
test['predictions'].values, opacity = 0.8, name='prediction')
layout = dict(
    title= "Prediction vs actual:",
```

```
xaxis=dict(
    rangeselector=dict(
        buttons=list([
            dict(count=1, label='1m', step='month',
            stepmode='backward'),
            dict(count=6, label='6m', step='month',
            stepmode='backward'),
            dict(count=12, label='12m', step='month',
            stepmode='backward'),
            dict(step='all')
        ])
    ),
    rangeslider=dict(visible = True),
    type='date'
    )
)
fig = dict(data= [trace0,trace1], layout=layout)
iplot(fig)
```

Figure 4-10 shows the plot of predictions vs. actuals for the test set.

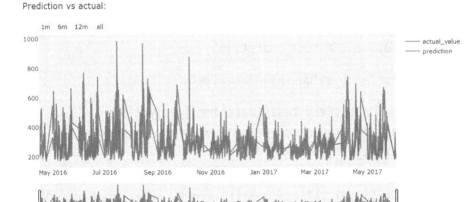

Figure 4-10. *Output*

Step 5-4. Compare performance in the validation set.

Let's check the best performance model for the validation set.

```
models=[xg_reg,lgb_reg,regr]
model_name=['XGBoost','LightGBM','RandomForest']
model_RMSE=[]
model_predictions=[]
for item in models:
    base_accuracy,predictions,RMSE=evaluate(item,X_pred,y_pred)
    model_RMSE.append(RMSE)
    model_predictions.append(predictions)
r=model_RMSE.index(min(model_RMSE))
best_model_predictions=model_predictions[r]
best_model_name=model_name[r]
best_model=models[r]
```

The output is as follows.

```
Model Performance
Average Error: 54.9496 degrees.
Accuracy = 81.69%.
RMSE = 81.81
Model Performance
Average Error: 43.6196 degrees.
Accuracy = 85.48%.
RMSE = 67.90
Model Performance
Average Error: 46.8309 degrees.
Accuracy = 83.58%.
RMSE = 71.68

print('Best Model:')
print(best_model_name)
```

```
print('Model Object:')
print(best_model)
print('Predictions:')
print(best_model_predictions)
```

Figure 4-11 shows the output of the best model, object and predictions for the validation set.

The output is as follows.

```
Best Model:
LightGBM
Model Object:
LGBMRegressor(random_state=42)
Predictions:
[192.02849511 193.2968421  237.88839221 221.5189054  212.80355811
 206.80779746 207.37546971 207.14007037 208.5919119  205.05943497
 202.49199157 205.05943497 206.73860635 203.37080023 207.00436673
 ???? ?????????  ??? ????????? ???? ?????????? ??? ????????? ??? ????????
```

Figure 4-11. Output

Among the three regressor models, LightGBM performs the best in both test sets. It has the lowest RMSE (67).

Let's use the LightGBM model predictions.

Step 5-5. Plot the LightGBM model prediction against actuals in the validation set.

Let's plot predictions vs. actuals charts using Plotly for the validation set.

```
#Plot timeseries
y_pred=pd.DataFrame(y_pred)

y_pred['predictions']=best_model_predictions

X_pred['datetime']=pd.to_datetime(X_pred[['year','month','day',
'hour']])
```

```python
y_pred['datetime']=X_pred['datetime']

y_pred=y_pred.sort_values(by='datetime')

trace0 = go.Scatter(x=y_pred['datetime'].astype(str),
y=y_pred['electricity_consumption'].values, opacity = 0.8,
name='actual_value')
trace1 = go.Scatter(x=y_pred['datetime'].astype(str), y=y_
pred['predictions'].values, opacity = 0.8, name='prediction')
layout = dict(
    title= "Prediction vs actual:",
    xaxis=dict(
        rangeselector=dict(
            buttons=list([
                dict(count=1, label='1m', step='month',
                stepmode='backward'),
                dict(count=6, label='6m', step='month',
                stepmode='backward'),
                dict(count=12, label='12m', step='month',
                stepmode='backward'),
                dict(step='all')
            ])
        ),
        rangeslider=dict(visible = True),
        type='date'
    )
)
fig = dict(data= [trace0,trace1], layout=layout)
iplot(fig)
```

Figure 4-12 shows the plot of predictions vs. actuals for the validation set.

Prediction vs actual:

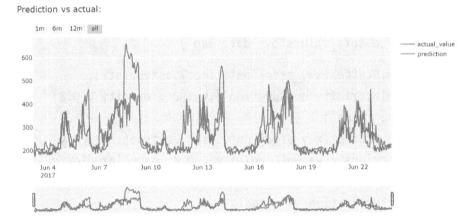

Figure 4-12. *Output*

Looking at both charts, the predictions are decent in the LightGBM model.

CHAPTER 5

Deep Learning–based Time Series Forecasting

Deep learning methods offer much promise for time series forecasting, such as automatic learning of temporal dependence and automatic processing of temporal structures such as trends and seasonality.

Due to the increasing availability of data and computing power in recent years, Deep learning has become an essential part of the new generation of time series forecasting models and has achieved excellent results.

While in classical machine learning models—such as autoregressive models (AR) or exponential smoothing—feature engineering is done manually, some parameters are often optimized with domain knowledge in mind. Deep learning models learn features and dynamics only and directly from data. Thanks to this, they speed up the data preparation process and can comprehensively learn more complex data patterns.

This chapter includes the following topics.

> Recipe 5-1. Time Series Forecasting Using LSTM
>
> Recipe 5-2. Multivariate Time Series Forecasting Using the GRU Model

© Akshay R Kulkarni, Adarsha Shivananda, Anoosh Kulkarni, V Adithya Krishnan 2023
A. R. Kulkarni et al., *Time Series Algorithms Recipes*,
https://doi.org/10.1007/978-1-4842-8978-5_5

Recipe 5-3. Time Series Forecasting Using
NeuralProphet

Recipe 5-4. Time Series Forecasting Using RNN

Recipe 5-1. Time Series Forecasting Using LSTM

Problem

You want to load the time series data and forecast using LSTM.

Solution

It can be easily achieved by using the built-in method defined in Keras.

How It Works

The following steps use LSTM to read the data and forecast.

Step 1-1. Import the required libraries.

```
import numpy as np
import pandas as pd
import matplotlib.pyplot as plt
import tensorflow as tf
import sklearn.preprocessing
from sklearn.metrics import r2_score
from keras.layers import Dense,Dropout,SimpleRNN,LSTM
from keras.models import Sequential
```

Step 1-2. Use DOM_hourly.csv data for analysis.

Let's use the DOM_hourly data, which is the Dominion Energy (DOM) time series data, to measure estimated energy consumption in megawatts (MW).

```
file_path ='./data/DOM_hourly.csv'
```

Step 1-3. Read the data.

```
data = pd.read_csv(file_path, index_col='Datetime', parse_
dates=['Datetime'])
data.head()
```

Figure 5-1 shows the head of the dataframe.

	DOM_MW
Datetime	
2005-12-31 01:00:00	9389.0
2005-12-31 02:00:00	9070.0
2005-12-31 03:00:00	9001.0
2005-12-31 04:00:00	9042.0
2005-12-31 05:00:00	9132.0

Figure 5-1. *Output*

Step 1-4. Check for missing data.

```
data.isna().sum()
```

The output is as follows.

```
DOM_MW    0
dtype: int64
```

Step 1-5. Plot the time series data.

```
data.plot(figsize=(16,4),legend=True)
```

```
plt.title('DOM hourly power consumption data - BEFORE
NORMALIZATION')
plt.show()
```

Figure 5-2 shows the time series plot.

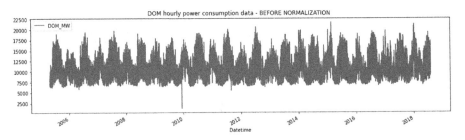

Figure 5-2. *Output*

Step 1-6. Write a function to normalize the data.

```
def normalize_fn(data):

    scaler_object = sklearn.preprocessing.MinMaxScaler()
    data['DOM_MW']=scaler_object.fit_transform(data['DOM_MW'].
    values.reshape(-1,1))
    return data
```

Step 1-7. Call the normalize_fn function.

```
data_norm = normalize_fn(data)
data_norm.head()
```

Figure 5-3 shows the output dataframe.

	DOM_MW
Datetime	
2005-12-31 01:00:00	0.398863
2005-12-31 02:00:00	0.383224
2005-12-31 03:00:00	0.379841
2005-12-31 04:00:00	0.381851
2005-12-31 05:00:00	0.386263

Figure 5-3. *Output*

Step 1-7. Plot the data after normalization.

```
data_norm.plot(figsize=(16,4),legend=True)

plt.title('DOM hourly power consumption data - AFTER
NORMALIZATION')

plt.show()
```

Step 1-8. Create a function to perform data preparation and train-test split.

```
def data_prep(data, length):
    X = []
    y = []

    for i in range(length, len(data)):
        X.append(data.iloc[i - length: i, 0])
        y.append(data.iloc[i, 0])

    # train-test split
    # training contains first 110000 days and test contains the
    remaining 6189 days
    train_X = X[:110000]
```

```
train_y = y[:110000]

test_X = X[110000:]
test_y = y[110000:]

# converting to numpy array
train_X = np.array(train_X)
train_y = np.array(train_y)

test_X = np.array(test_X)
test_y = np.array(test_y)

# reshaping data to required format to input to RNN,
LSTM models
train_X = np.reshape(train_X, (110000, length, 1))
test_X = np.reshape(test_X, (test_X.shape[0], length, 1))

return [train_X, train_y, test_X, test_y]
```

Step 1-9. Call the data_prep function.

```
sequence_length = 20
train_X, train_y, test_X, test_y = data_prep(data,
sequence_length)

print('train_X.shape = ',train_X.shape)
print('train_y.shape = ', train_y.shape)
print('test_X.shape = ', test_X.shape)
print('test_y.shape = ',test_y.shape)
```

Step 1-10. Initialize the LSTM model.

```
model = Sequential()
model.add(LSTM(40,activation="tanh",return_sequences=True,
input_shape=(train_X.shape[1],1)))
```

```
model.add(Dropout(0.15))
model.add(LSTM(40,activation="tanh",return_sequences=True))
model.add(Dropout(0.15))
model.add(LSTM(40,activation="tanh",return_sequences=False))
model.add(Dropout(0.15))
model.add(Dense(1))
```

Step 1-11. Create the model summary.

```
model.summary()
```

Figure 5-4 shows the model summary.

```
Model: "sequential_1"
```

Layer (type)	Output Shape	Param #
lstm (LSTM)	(None, 20, 40)	6720
dropout_3 (Dropout)	(None, 20, 40)	0
lstm_1 (LSTM)	(None, 20, 40)	12960
dropout_4 (Dropout)	(None, 20, 40)	0
lstm_2 (LSTM)	(None, 40)	12960
dropout_5 (Dropout)	(None, 40)	0
dense_1 (Dense)	(None, 1)	41

```
Total params: 32,681
Trainable params: 32,681
Non-trainable params: 0
```

Figure 5-4. *Output*

Step 1-12. Fit the model.

```
model.compile(optimizer="adam",loss="MSE")
model.fit(train_X, train_y, epochs=10, batch_size=1000)
```

Figure 5-5 shows the epochs 1 to 10.

```
Epoch 1/10
110/110 [==============================] - 26s 193ms/step - loss: 0.0211
Epoch 2/10
110/110 [==============================] - 20s 185ms/step - loss: 0.0119
Epoch 3/10
110/110 [==============================] - 21s 191ms/step - loss: 0.0080
Epoch 4/10
110/110 [==============================] - 21s 186ms/step - loss: 0.0047
Epoch 5/10
110/110 [==============================] - 20s 185ms/step - loss: 0.0037
Epoch 6/10
110/110 [==============================] - 21s 194ms/step - loss: 0.0030
Epoch 7/10
110/110 [==============================] - 20s 185ms/step - loss: 0.0026
Epoch 8/10
110/110 [==============================] - 21s 192ms/step - loss: 0.0022
Epoch 9/10
110/110 [==============================] - 20s 185ms/step - loss: 0.0020
Epoch 10/10
110/110 [==============================] - 21s 191ms/step - loss: 0.0018
```

```
<keras.callbacks.History at 0x7f01d4e38810>
```

Figure 5-5. *Output*

Step 1-13. Make the model predictions and print the score.

```
predictions = model.predict(test_X)
score = r2_score(test_y,predictions)
print("R-Squared Score of LSTM model",score)
```

The output is as follows.

```
R-Squared Score of LSTM model =  0.94996673239313
```

Step 1-14. Write a function to plot the predictions.

```
def plotting_actual_vs_pred(y_test, y_pred, title):
    plt.figure(figsize=(16, 4))
    plt.plot(y_test, color='blue', label='Actual power
    consumption data')
    plt.plot(y_pred, alpha=0.7, color='orange',
    label='Predicted power consumption data')
    plt.title(title)
    plt.xlabel('Time')
    plt.ylabel('Normalized power consumption scale')
    plt.legend()
    plt.show()
```

Step 1-15. Call the plotting_actual_vs_pred function.

```
plotting_actual_vs_pred(test_y, predictions, "Predictions made
by LSTM model")
```

Figure 5-6 shows the actual vs. prediction plot.

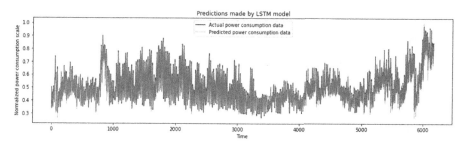

Figure 5-6. *Output*

Recipe 5-2. Multivariate Time Series Forecasting Using the GRU Model

Problem

You want to load time series data with multiple targets and forecast them using GRU.

Solution

It can be easily achieved using a built-in method defined in Keras.

How It Works

The following steps use GRU to read the data and forecast.

Step 2-1. Import the required libraries.

```
#import all the required libraries

import tensorflow as tf
from tensorflow import keras
import numpy as np
import pandas as pd
import matplotlib.pyplot as plt
import seaborn as sns
```

Step 2-2. Read the data.

Download the data from the Git link.

The following code reads the data.

```
#read data

train_data = pd.read_csv("../input/daily-climate-time-series-
data/DailyDelhiClimateTrain.csv",index_col=0)
# Display dimensions of dataframe
print(train_data.shape)
print(train_data.info())
```

Figure 5-7 shows the output.

```
(1462, 4)
<class 'pandas.core.frame.DataFrame'>
Index: 1462 entries, 2013-01-01 to 2017-01-01
Data columns (total 4 columns):
 #   Column         Non-Null Count  Dtype
---  ------         --------------  -----
 0   meantemp       1462 non-null   float64
 1   humidity       1462 non-null   float64
 2   wind_speed     1462 non-null   float64
 3   meanpressure   1462 non-null   float64
dtypes: float64(4)
memory usage: 57.1+ KB
None
```

Figure 5-7. *Output*

Along with the date column, the other four features that need to be forecasted are mean temp, humidity, wind speed, and mean pressure.

Step 2-3. Analyze the data.

Let's check the random sample records for the dataset and perform univariate analysis to see the basic stats of each column.

```
# sample records
print("-----")
print("Original dataset  : \n",train_data.sample(10))
```

```
# Univariate analysis
print("------
-------")
# Display statistics for numeric columns
print(train_data.describe())
```

Figure 5-8 shows the sample and describes the dataframe.

```
--------------------------------------------------------------------
Original dataset  :
               meantemp  humidity  wind_speed  meanpressure
date
2014-04-07      27.250    47.625    12.962500   1009.625000
2014-12-16      17.750    72.375     7.425000   1017.375000
2015-02-25      23.125    58.625     6.500000   1008.500000
2014-06-30      32.000    67.750     3.250000    997.500000
2013-12-21      14.750    94.000     0.462500   1017.000000
2015-01-30      12.750    56.125    12.037500   1020.375000
2015-02-02      17.375    63.875    11.812500   1017.500000
2015-04-20      32.875    37.875     7.187500   1005.750000
2014-02-12      13.250    67.000     9.262500   1013.500000
2016-05-22      36.800    44.800     6.553333    996.466667

--------------------------------------------------------------------
             meantemp      humidity    wind_speed   meanpressure
count     1462.000000   1462.000000   1462.000000    1462.000000
mean        25.495521     60.771702      6.802209    1011.104548
std          7.348103     16.769652      4.561602     180.231668
min          6.000000     13.428571      0.000000      -3.041667
25%         18.857143     50.375000      3.475000    1001.580357
50%         27.714286     62.625000      6.221667    1008.563492
75%         31.305804     72.218750      9.238235    1014.944901
max         38.714286    100.000000     42.220000    7679.333333
```

Figure 5-8. *Output*

The dataset consists of four columns against date so let's plot them and try to understand the trend.

```
#line plots to understand the trend
print("--------------------------------------------")
train_data.plot(figsize=(12,8),subplots=True)
```
Figure 5-9 shows the trend for all four variables.

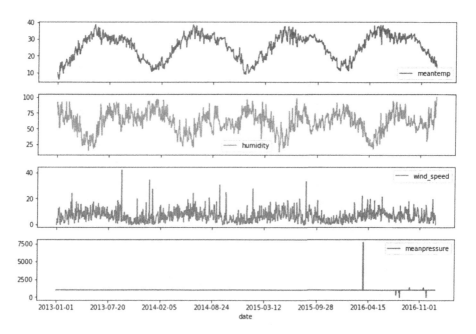

Figure 5-9. *Output*

Let's plot them all using a histogram to analyze the distribution.

```
#line plots to understand the trend

print("-------------")
train_data.plot(figsize=(12,8),subplots=True)
```

Figure 5-10 shows the distributions for all four variables.

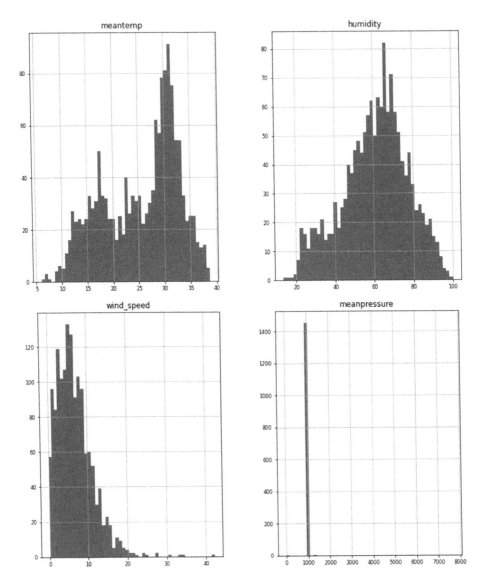

Figure 5-10. *Output*

Step 2-4. Preprocess the data.

After analyzing the data, let's ensure there are no nulls present in the data.

```
#check missing values
```

```
print("null values : \n",train_data.isnull().sum())
sns.heatmap(train_data.isnull(), cbar=False, yticklabels=False,
cmap="viridis")
```

Figure 5-11 shows the "check missing values" output.

The output is as follows.

```
null values :
 meantemp          0
humidity          0
wind_speed        0
meanpressure      0
dtype: int64

<AxesSubplot:ylabel='date'>
```

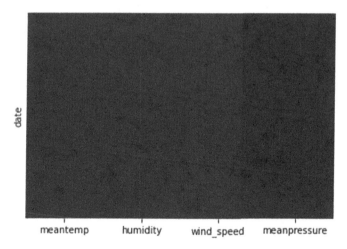

Figure 5-11. *Output*

There are no nulls present. Let's select the features and ensure that float is the datatype for all columns.

```
# We choose a specific feature (features). In this example,
my_dataset = train_data[["meantemp",'humidity','wind_
speed','meanpressure']]

print("Our new dataset : \n",my_dataset.sample(5))

# change datatype
print("-------------")
# ensure all data is float
my_dataset = my_dataset.astype("float32")
values      = my_dataset.values
print("values : \n",values)
```

Figure 5-12 shows the sample dataset.

```
Our new dataset :
               meantemp    humidity   wind_speed   meanpressure
date
2014-10-11   25.714286   57.142857     6.937500    1008.250000
2016-12-17   17.500000   63.388889     6.731579    1016.947368
2014-04-06   30.125000   45.250000     6.712500    1007.500000
2015-12-07   17.625000   76.875000     2.312500    1017.250000
2015-04-26   28.750000   51.125000    15.737500    1008.000000
-------------------------------------------------------------
values :
 [[  10.          84.5           0.        1015.6667   ]
 [   7.4         92.            2.98      1017.8      ]
 [   7.1666665   87.            4.633333  1018.6667   ]
 ...
 [  14.095238    89.666664      6.266667  1017.9048   ]
 [  15.052631    87.            7.325     1016.1      ]
 [  10.         100.            0.        1016.      ]]
```

Figure 5-12. *Output*

Now, let's normalize the features using MinMaxScaler.

```
# # normalize features
from sklearn.preprocessing import MinMaxScaler
scaler = MinMaxScaler(feature_range=(0, 1))
scaled = scaler.fit_transform(values)
print("scaled : \n",scaled)print("values : \n",values)
```

The output is as follows.

```
scaled :
 [[0.12227073 0.8209571  0.         0.1326033 ]
 [0.04279476 0.9075908  0.07058267 0.13288099]
 [0.03566229 0.849835   0.10974261 0.1329938 ]
 ...
 [0.24745268 0.88063806 0.14842887 0.13289464]
 [0.276718   0.849835   0.17349596 0.1326597 ]
 [0.12227073 1.0000001  0.         0.1326467 ]]
```

Now let's convert the series into supervised learning.

```
## convert series to supervised learning
def series_to_supervised(data, n_in=1, n_out=1, dropnan=True):
    n_vars = 1 if type(data) is list else data.shape[1]
    df = pd.DataFrame(data)
    cols, names = list(), list()
    # input sequence (t-n, ... t-1)
    for i in range(n_in, 0, -1):
        cols.append(df.shift(i))
        names += [("var%d(t-%d)" % (j+1, i)) for j in
range(n_vars)]
    # forecast sequence (t, t+1, ... t+n)
    for i in range(0, n_out):
        cols.append(df.shift(-i))
        if i == 0:
```

```
            names += [("var%d(t)" % (j+1)) for j in
            range(n_vars)]
        else:
            names += [("var%d(t+%d)" % (j+1, i)) for j in
            range(n_vars)]
    # put it all together
    agg = pd.concat(cols, axis=1)
    agg.columns = names
    # drop rows with NaN values
    if dropnan:
        agg.dropna(inplace=True)
    return agg

# call the function


# frame as supervised learning
# reshape into X=t and Y=t+1
i_in  = 100 # past observations
n_out = 1 # future observations
reframed = series_to_supervised(scaled, i_in, n_out)
print("Represent the dataset as a supervised learning problem :
\n",reframed.head(10))
```

The output is as follows.

```
epresent the dataset as a supervised learning problem :
      var1(t-100)  var2(t-100)  var3(t-100)  var4(t-100)  var1(t-99) \
100     0.122271     0.820957     0.000000     0.132603     0.042795
101     0.042795     0.907591     0.070583     0.132881     0.035662
102     0.035662     0.849835     0.109743     0.132994     0.081514
103     0.081514     0.668867     0.029212     0.132799     0.000000
104     0.000000     0.847910     0.087636     0.132712     0.030568
105     0.030568     0.801320     0.035054     0.132907     0.030568
```

106	0.030568	0.752805	0.149218	0.133167	0.087336
107	0.087336	0.580858	0.169182	0.133000	0.244541
108	0.244541	0.436881	0.296068	0.132777	0.152838
109	0.152838	0.561056	0.175272	0.132603	0.296943

	var2(t-99)	var3(t-99)	var4(t-99)	var1(t-98)	var2(t-98)	... \
100	0.907591	0.070583	0.132881	0.035662	0.849835	...
101	0.849835	0.109743	0.132994	0.081514	0.668867	...
102	0.668867	0.029212	0.132799	0.000000	0.847910	...
103	0.847910	0.087636	0.132712	0.030568	0.801320	...
104	0.801320	0.035054	0.132907	0.030568	0.752805	...
105	0.752805	0.149218	0.133167	0.087336	0.580858	...
106	0.580858	0.169182	0.133000	0.244541	0.436881	...
107	0.436881	0.296068	0.132777	0.152838	0.561056	...
108	0.561056	0.175272	0.132603	0.296943	0.437294	...
109	0.437294	0.250389	0.132665	0.244541	0.699670	...

	var3(t-2)	var4(t-2)	var1(t-1)	var2(t-1)	var3(t-1)	var4(t-1) \
100	0.204879	0.131258	0.733624	0.168317	0.144481	0.131438
101	0.144481	0.131438	0.733624	0.124422	0.184273	0.131397
102	0.184273	0.131397	0.698690	0.221122	0.150233	0.131550
103	0.150233	0.131550	0.739738	0.182178	0.236855	0.131657
104	0.236855	0.131657	0.680131	0.299711	0.153659	0.131475
105	0.153659	0.131475	0.680131	0.322814	0.148034	0.131068
106	0.148034	0.131068	0.798581	0.129332	0.246921	0.130564
107	0.246921	0.130564	0.709170	0.124422	0.157745	0.130850
108	0.157745	0.130850	0.742358	0.194719	0.112675	0.130936
109	0.112675	0.130936	0.681223	0.206271	0.094065	0.130899

	var1(t)	var2(t)	var3(t)	var4(t)
100	0.733624	0.124422	0.184273	0.131397
101	0.698690	0.221122	0.150233	0.131550
102	0.739738	0.182178	0.236855	0.131657

```
103  0.680131  0.299711  0.153659  0.131475
104  0.680131  0.322814  0.148034  0.131068
105  0.798581  0.129332  0.246921  0.130564
106  0.709170  0.124422  0.157745  0.130850
107  0.742358  0.194719  0.112675  0.130936
108  0.681223  0.206271  0.094065  0.130899
109  0.752729  0.179868  0.180898  0.131003

[10 rows x 404 columns]
```

Step 2-5. Do a train-test split.

Let's split the data into train and test sets, using the test set to validate the model.

```
# # split into train and test sets
# convert an array of values into a dataset matrix
values_spl = reframed.values
train_size = int(len(values_spl) * 0.80)
test_size  = len(values_spl) - train_size
train, test = values_spl[0:train_size,:], values_spl
[train_size:len(values_spl),:]

print("len train and test : ",len(train), "  ", len(test))

print("-------------")
# split into input and outputs
X_train, y_train = train[:, :-4], train[:, -4:]
X_test, y_test   = test[:, :-4],  test[:, -4:]

print("X_train shape : ",X_train.shape," y_train shape : ",
y_train.shape)
print("X_test shape  : ",X_test.shape, " y_test shape  : ",
y_test.shape)
```

```
print("-------------")
# reshape input to be 3D [samples, timesteps, features]
# The LSTM network expects the input data (X) to be
provided with
# a specific array structure in the form of: [samples, time
steps, features].
# Currently, our data is in the form: [samples, features]
X_train = X_train.reshape((X_train.shape[0], 1, X_train.
shape[1]))
X_test  = X_test.reshape((X_test.shape[0], 1, X_test.shape[1]))

print("X_train shape 3D : ",X_train.shape," y_train shape :
",y_train.shape)
print("X_test shape  3D : ",X_test.shape, " y_test shape  :
",y_test.shape)
```

The output is as follows.

```
len train and test :  1089     273

-------------
X_train shape :  (1089, 400)  y_train shape :  (1089, 4)
X_test shape  :  (273, 400)  y_test shape  :  (273, 4)
-------------

X_train shape 3D :  (1089, 1, 400)  y_train shape :  (1089, 4)
X_test shape  3D :  (273, 1, 400)  y_test shape  :  (273, 4)
```

Step 2-6. Build the model.

Let's build the GRU model that is defined in Keras.

First, let's import and define all the layers.

```
# #import and define the layers

model = keras.models.Sequential()
```

```
model.add(keras.layers.GRU(64, return_sequences=True,
activation="relu",
            kernel_initializer="he_normal", recurrent_
            initializer="he_normal",
            dropout=0.15, recurrent_dropout=0.15,
                        input_shape=(X_train.shape[1], X_train.
                        shape[2]) ))
model.add(keras.layers.GRU(32,return_sequences=True,
activation="relu", kernel_initializer="he_normal",
            recurrent_initializer="he_normal", dropout=0.15,
            recurrent_dropout=0.15 ))
model.add(keras.layers.GRU(8, activation="relu", kernel_
initializer="he_normal",
            recurrent_initializer="he_normal", dropout=0.15,
            recurrent_dropout=0.15 ))
model.add(keras.layers.Dense(4, activation="relu"))

print(model.summary())
```

The output is as follows.

```
Model: "sequential_2"
```

Layer (type)	Output Shape	Param #
gru_6 (GRU)	(None, 1, 64)	89472
gru_7 (GRU)	(None, 1, 32)	9408
gru_8 (GRU)	(None, 8)	1008
dense_2 (Dense)	(None, 4)	36

Total params: 99,924
Trainable params: 99,924
Non-trainable params: 0

None

Let's plot and check.

#plot model

```
from tensorflow.keras.utils import plot_model
plot_model(model, show_shapes=True)
```

Figure 5-13 shows the model plot.

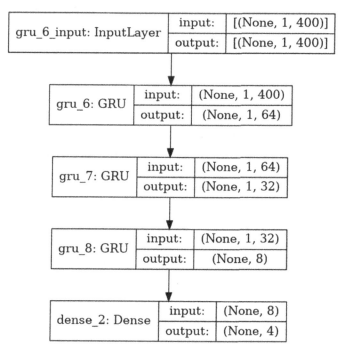

Figure 5-13. *Output*

Let's compile and fit the model.

```
## Compiling the model
optimizer = keras.optimizers.Adam(learning_rate=0.01,
beta_1=0.9, beta_2=0.999)
model.compile(loss="mean_squared_error", optimizer=optimizer,
metrics=["mse","mae"])

# Learning rate scheduling
lr_scheduler = keras.callbacks.ReduceLROnPlateau(fact
or=0.00001, patience=3, monitor="val_loss", min_lr=0.00000001)

# Training and evaluating the model
history = model.fit(X_train, y_train, epochs=100, batch_
size=64, validation_split=0.2,
                    callbacks=[lr_scheduler])
```

The output is as follows.

```
Epoch 1/100
14/14 [==============================] - 10s 88ms/step - loss:
0.1164 - mse: 0.1164 - mae: 0.2643 - val_loss: 0.0367 -
val_mse: 0.0367 - val_mae: 0.1475
Epoch 2/100
14/14 [==============================] - 0s 14ms/step - loss:
0.0437 - mse: 0.0437 - mae: 0.1532 - val_loss: 0.0142 -
val_mse: 0.0142 - val_mae: 0.0913
Epoch 3/100
14/14 [==============================] - 0s 15ms/step - loss:
0.0236 - mse: 0.0236 - mae: 0.1087 - val_loss: 0.0135 -
val_mse: 0.0135 - val_mae: 0.0789
Epoch 4/100
```

```
14/14 [==============================] - 0s 14ms/step - loss:
0.0192 - mse: 0.0192 - mae: 0.0962 - val_loss: 0.0121 -
val_mse: 0.0121 - val_mae: 0.0789
Epoch 5/100
14/14 [==============================] - 0s 14ms/step - loss:
0.0170 - mse: 0.0170 - mae: 0.0906 - val_loss: 0.0102 -
val_mse: 0.0102 - val_mae: 0.0691
Epoch 6/100
14/14 [==============================] - 0s 14ms/step - loss:
0.0153 - mse: 0.0153 - mae: 0.0850 - val_loss: 0.0087 -
val_mse: 0.0087 - val_mae: 0.0640
```

Step 2-7. Evaluate and predict the model.

Let's plot evaluation metrics like MSE and MAE (refer to Chapter 4 for the definitions).

```
# # plot the learning curves
pd.DataFrame(history.history).plot(figsize=(8, 5))
plt.grid(True)
plt.gca().set_ylim(0, 1) # set the vertical range to [0-1]
plt.show()
```

Figure 5-14 shows the model evaluation metrics plot.

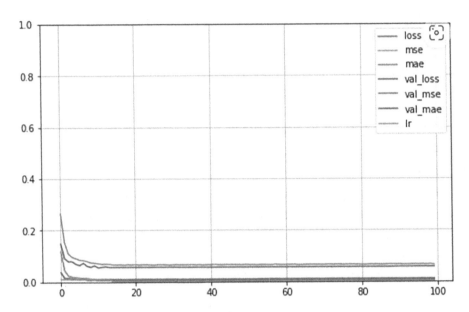

Figure 5-14. *Output*

```
print("-------------")
# Evaluate the model
model_evaluate = model.evaluate(X_test, y_test)
print("Loss                  : ",model_evaluate[0])
print("Mean Squared Error    : ",model_evaluate[1])
print("Mean Absolute Error   : ",model_evaluate[2])

# make predictions
trainPredict = model.predict(X_train)
testPredict  = model.predict(X_test)
print("trainPredict : ",trainPredict.shape)
print("testPredict  : ",testPredict.shape)

print(trainPredict)

testPredict = scaler.inverse_transform(testPredict)
```

```
print(testPredict.shape)
print(y_test.shape)

y_test=scaler.inverse_transform(y_test)
```

The output is as follows.

```
-------------------------------------------------------------------
9/9 [==============================] - 0s 3ms/step - loss: 0.0084 - mse: 0.0084 - mae: 0.0652
Loss                  :  0.008361274376511574
Mean Squared Error    :  0.008361274376511574
Mean Absolute Error   :  0.06517541408538818
```

```
trainPredict :  (1089, 4)
testPredict  :  (273, 4)

array([[0.794252  , 0.29386416, 0.16390276, 0.13381857],
       [0.80810547, 0.28626317, 0.16907552, 0.1337166 ],
       [0.8088531 , 0.28755552, 0.1696803 , 0.13367665],
       ...,
       [0.67830986, 0.43075496, 0.139501  , 0.13334922],
       [0.69778967, 0.4183182 , 0.14170927, 0.13308588],
       [0.7168243 , 0.40688953, 0.14848372, 0.13296095]],
      dtype=float32)

(273, 4)
(273, 4)
```

Now that there are predictions for all four features, let's plot a line chart against the actuals.

1. Plot the mean temp.

    ```
    ##plot for meantemp

    plt.plot(testPredict[:,0], color="blue",
            label="Predict meantemp ", linewidth=2)
    ```

```
plt.plot(y_test[:,0], color="red",
        label="Actual meantemp ", linewidth=2)
```

```
plt.legend()
# Show the major grid lines with dark grey lines
plt.grid(visible=True, which="major", color="#666666",
linestyle="-")
# Show the minor grid lines with very faint and almost
transparent grey lines
plt.minorticks_on()
plt.grid(visible=True, which="minor", color="#999999",
linestyle="-", alpha=0.2)
```

```
plt.show()
```

Figure 5-15 shows the predictions vs. actuals plot for the meantemp column.

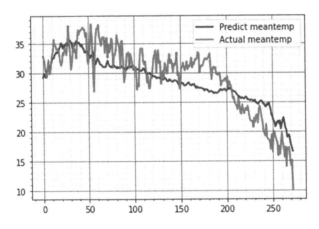

Figure 5-15. *Output*

2. Plot the humidity.

    ```
    ## #plot for humidity
    ```

```
plt.plot(testPredict[:,1], color="blue",
        label="Predict humidity", linewidth=2)
plt.plot(y_test[:,1], color="red",
        label="Actual humidity", linewidth=2)
plt.legend()

# Show the major grid lines with dark grey lines
plt.grid(visible=True, which="major", color="#666666",
linestyle="-")
# Show the minor grid lines with very faint and almost
transparent grey lines
plt.minorticks_on()
plt.grid(visible=True, which="minor", color="#999999",
linestyle="-", alpha=0.2)
plt.show()
```

Figure 5-16 shows the predictions vs actuals plot for Humidity column

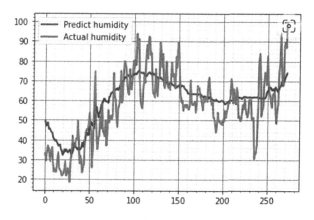

Figure 5-16. *Output*

3. Plot the wind speed.

```
# #plot for windspeed
```

```
plt.plot(testPredict[:,2], color="blue",
        label="predict wind_speed", linewidth=2)
plt.plot(y_test[:,2], color="red",
        label="Actual wind_speed", linewidth=2)
plt.legend()

# Show the major grid lines with dark grey lines
plt.grid(visible=True, which="major", color="#666666",
linestyle="-")
# Show the minor grid lines with very faint and almost
transparent grey lines
plt.minorticks_on()
plt.grid(visible=True, which="minor", color="#999999",
linestyle="-", alpha=0.2)

plt.show()
```

Figure 5-17 shows the predictions vs. actuals plot for the wind_speed column.

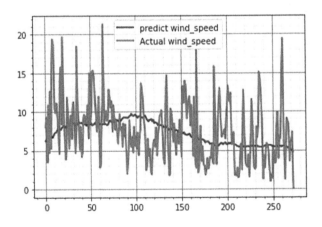

Figure 5-17. *Output*

4. Plot the mean pressure.

```
# plot for meanpressure

plt.plot(testPredict[:,3], color="blue",
        label="predict meanpressure", linewidth=4)
plt.plot(y_test[:,3], color="red",
        label="Actual meanpressure", linewidth=4)

plt.legend()

# Show the major grid lines with dark grey lines
plt.grid(visible=True, which="major", color="#666666",
linestyle="-")
# Show the minor grid lines with very faint and almost
transparent grey lines
plt.minorticks_on()
plt.grid(visible=True, which="minor", color="#999999",
linestyle="-", alpha=0.2)

plt.show()
```

Figure 5-18 shows the predictions vs. actuals plot for the meanpressure column.

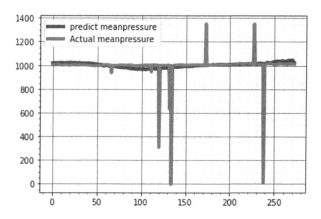

Figure 5-18. *Output*

Recipe 5-3. Time Series Forecasting Using NeuralProphet

Problem

You want to load the univariate time series data and forecast using NeuralProphet.

Solution

It can be easily achieved by using the built-in package.

How It Works

The following steps use NeuralProphet to read the data and forecast.

Step 3-1. Import the required libraries.

```
#import all the required libraries

import pandas as pd
from neuralprophet import NeuralProphet
import matplotlib.pyplot as plt
```

Step 3-2. Read the data.

Download the data from the Git link.

The following code reads the data. While reading the data, let's parse the date column.

```
#read data

df_np = pd.read_csv("./DailyDelhiClimateTrain.csv",
parse_dates=["date"])
```

Step 3-3. Preprocess the data.

Let's preprocess the data as per the NeuralProphet requirements.

Since it is univariate, select only one feature to forecast with the date column.

Also, NeuralProphet expects the date column to be named "ds" and the target column to be named "y".

```
#Data pre-process

df_np = df_np[["date", "meantemp"]]
df_np.rename(columns={"date": "ds", "meantemp": "y"},
inplace=True)
```

Step 3-4. Build the model and make predictions.

Let's initialize the NeuralProphet model and define all the arguments, including additional information. If you plan to use default variables, use model = NeuralProphet().

```
# model = NeuralProphet() if you're using default
variables below.
model = NeuralProphet(
    growth="linear",  # Determine trend types: 'linear',
'discontinuous', 'off'
    changepoints=None, # list of dates that may include change
points (None -> automatic )
    n_changepoints=5,
    changepoints_range=0.8,
    trend_reg=0,
    trend_reg_threshold=False,
    yearly_seasonality="auto",
    weekly_seasonality="auto",
    daily_seasonality="auto",
```

```
    seasonality_mode="additive",
    seasonality_reg=0,
    n_forecasts=1,
    n_lags=0,
    num_hidden_layers=0,
    d_hidden=None,      # Dimension of hidden layers of AR-Net
    learning_rate=None,
    epochs=40,
    loss_func="Huber",
    normalize="auto",   # Type of normalization ('minmax',
'standardize', 'soft', 'off')
    impute_missing=True,
)
```

Once the model is initialized, let's fit the model and make predictions.

```
#make predictions
metrics = model.fit(df_np, freq="D")
future = model.make_future_dataframe(df_np, periods=365, n_
historic_predictions=len(df_np))
forecast = model.predict(future)
```

Figure 5-19 shows the output of the model.

```
INFO - (NP.df_utils._infer_frequency) - Major frequency D corresponds to 99.932% of the data.
INFO - (NP.df_utils._infer_frequency) - Defined frequency is equal to major frequency - D
INFO - (NP.config.init_data_params) - Setting normalization to global as only one dataframe provided for training.
INFO - (NP.utils.set_auto_seasonalities) - Disabling daily seasonality. Run NeuralProphet with daily_seasonality=True to ove
rride this.
INFO - (NP.config.set_auto_batch_epoch) - Auto-set batch_size to 32

100% ████████████████████████  129/129 [00:00<00:00, 310.12it/s]

INFO - (NP.utils_torch.lr_range_test) - lr-range-test results: steep: 9.14E-01, min: 1.54E-01

100% ████████████████████████  129/129 [00:00<00:00, 311.50it/s]

INFO - (NP.utils_torch.lr_range_test) - lr-range-test results: steep: 9.14E-01, min: 1.54E-01
INFO - (NP.forecaster._init_train_loader) - lr-range-test selected learning rate: 4.37E-01
Epoch[40/40]: 100%|████████████| 40/40 [00:04<00:00,  9.95it/s, SmoothL1Loss=0.00238, MAE=1.6, RMSE=2.02, RegLoss=0]
INFO - (NP.df_utils._infer_frequency) - Major frequency D corresponds to 99.932% of the data.
INFO - (NP.df_utils._infer_frequency) - Defined frequency is equal to major frequency - D
WARNING - (py.warnings._showwarnmsg) - C:\Users\ashwi\anaconda3\lib\site-packages\neuralprophet\forecaster.py:2060: FutureWa
rning: The frame.append method is deprecated and will be removed from pandas in a future version. Use pandas.concat instead.
  df = df.append(future_df)

INFO - (NP.df_utils._infer_frequency) - Major frequency D corresponds to 99.945% of the data.
INFO - (NP.df_utils._infer_frequency) - Defined frequency is equal to major frequency - D
INFO - (NP.df_utils._infer_frequency) - Major frequency D corresponds to 99.945% of the data.
INFO - (NP.df_utils._infer_frequency) - Defined frequency is equal to major frequency - D
WARNING - (py.warnings._showwarnmsg) - C:\Users\ashwi\anaconda3\lib\site-packages\neuralprophet\forecaster.py:1406: FutureWa
rning: The frame.append method is deprecated and will be removed from pandas in a future version. Use pandas.concat instead.
  df = df.append(df_end_to_append)
```

Figure 5-19. *Output*

Now, let's plot the graph to see the forecasted values.

```
##forecast plot
fig, ax = plt.subplots(figsize=(14, 10))
model.plot(forecast, xlabel="Date", ylabel="Temp", ax=ax)
ax.set_title("Mean Temperature in Delhi", fontsize=28,
fontweight="bold")
```

Figure 5-20 shows the predictions and actuals plot.

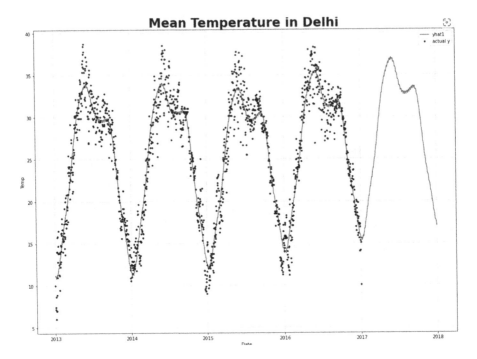

Figure 5-20. Output

The latest one-year forecast plot is shown. The forecasted values are from 2017 to 2018. The forecasted values resemble the historical, meaning the model has captured seasonality and the linear trend.

You can also plot the parameters.

```
##plotting model parameters
model.plot_parameters()
```

Figure 5-21 shows the model parameters plot.

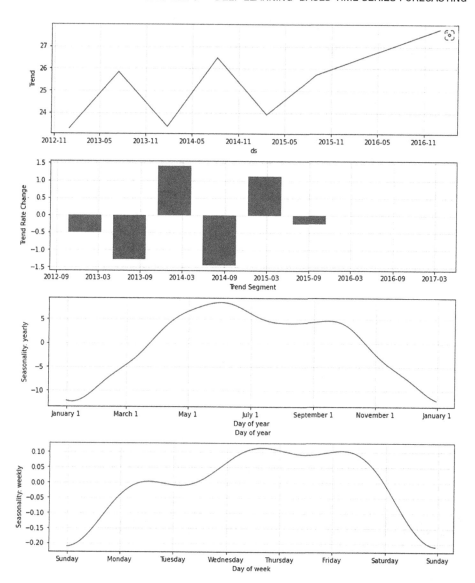

Figure 5-21. *Output*

The model loss using mean absolute error (MAE) is plotted as follows. You can also use the Smoothed L1 loss function.

###ploting Evaluation

```
fig, ax = plt.subplots(figsize=(14, 10))
ax.plot(metrics["MAE"], 'ob', linewidth=6)
ax.plot(metrics["RMSE"], '-r', linewidth=2)
```

Figure 5-22 shows the model evaluation plot.

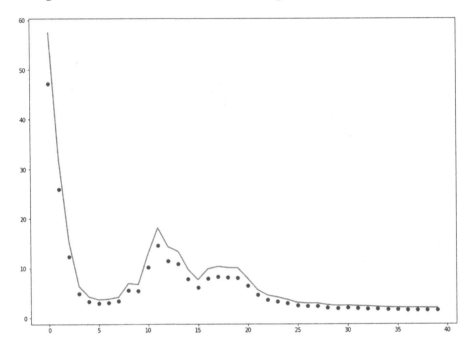

Figure 5-22. *Output*

Recipe 5-4. Time Series Forecasting Using RNN

Problem

You want to load the time series data and forecast using a *recurrent neural network* (RNN).

Solution

It can be easily achieved by using the built-in method defined in Keras.

How It Works

The following steps use RNN to read the data and forecast.

Steps 1-1 to 1-9 from Recipe 5-1 are also used for this recipe.

Step 4-1. Initialize the RNN model.

```
model = Sequential()
model.add(LSTM(40,activation="tanh",return_sequences=True,
input_shape=(train_X.shape[1],1)))
model.add(Dropout(0.15))
model.add(LSTM(40,activation="tanh",return_sequences=True))
model.add(Dropout(0.15))
model.add(LSTM(40,activation="tanh",return_sequences=False))
model.add(Dropout(0.15))
model.add(Dense(1))
```

Step 4-2. Create the model summary.

```
model.summary()
```

Figure 5-23 shows the model summary.

```
Model: "sequential"
```

Layer (type)	Output Shape	Param #
simple_rnn (SimpleRNN)	(None, 20, 40)	1680
dropout (Dropout)	(None, 20, 40)	0
simple_rnn_1 (SimpleRNN)	(None, 20, 40)	3240
dropout_1 (Dropout)	(None, 20, 40)	0
simple_rnn_2 (SimpleRNN)	(None, 40)	3240
dropout_2 (Dropout)	(None, 40)	0
dense (Dense)	(None, 1)	41

```
Total params: 8,201
Trainable params: 8,201
Non-trainable params: 0
```

Figure 5-23. *Output*

Step 4-3. Fit the model.

```
model.compile(optimizer="adam",loss="MSE")
model.fit(train_X, train_y, epochs=10, batch_size=1000)
```

Figure 5-24 shows epochs 1 to 10.

```
Epoch 1/10

2022-08-19 16:26:37.061384: I tensorflow/compiler/mlir/mlir_graph_optimiz
s are enabled (registered 2)

110/110 [==============================] - 10s 73ms/step - loss: 0.0820
Epoch 2/10
110/110 [==============================] - 9s 81ms/step - loss: 0.0178
Epoch 3/10
110/110 [==============================] - 8s 74ms/step - loss: 0.0096
Epoch 4/10
110/110 [==============================] - 8s 73ms/step - loss: 0.0065
Epoch 5/10
110/110 [==============================] - 8s 74ms/step - loss: 0.0050
Epoch 6/10
110/110 [==============================] - 8s 73ms/step - loss: 0.0040
Epoch 7/10
110/110 [==============================] - 9s 81ms/step - loss: 0.0035
Epoch 8/10
110/110 [==============================] - 8s 73ms/step - loss: 0.0030
Epoch 9/10
110/110 [==============================] - 8s 74ms/step - loss: 0.0027
Epoch 10/10
110/110 [==============================] - 8s 75ms/step - loss: 0.0024

<keras.callbacks.History at 0x7f01d517fc50>
```

Figure 5-24. *Output*

Step 4-4. Make the model predictions and print the score.

```
predictions = model.predict(test_X)
score = r2_score(test_y,predictions)
print("R-Squared Score of RNN model = ",score)
```

The output is as follows.

```
R-Squared Score of RNN model =  0.9466957722475382
```

Step 4-5 is the same as step 1-14 from Recipe 5-1.

Step 4-6. Call the plotting_actual_vs_pred function.

```
plotting_actual_vs_pred(test_y, predictions, "Predictions made
by simple RNN model")
```

Figure 5-25 shows the actual vs. prediction plot.

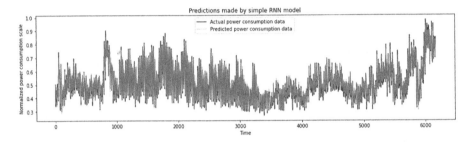

Figure 5-25. *Output*

Index

A

Additive model decomposition
adding components, 21
changes, 21
components separation, 24
libraries, loading, 22
plotting, 23
seasonality, 22
statsmodel library, 22
time series, 23, 24
turnover data, 22
Augmented Dickey-Fuller (ADF)
test, 40, 49, 90
Autoregressive (AR) models
actuals *vs.* predictions, 42, 43
autocorrelation
function, 40, 41
AutoReg function, 38
calling/fitting, 41
forecast, 38
lagged values, 38
libraries, importing, 39
plotting, 39
predictions, 42
stationarity, 40
summary, 41, 42
training/test data, 41

Autoregressive integrated moving
average (ARIMA) model
Auto Correlation Function/
Partial Auto Correlation
Function values, 50
ADF test, 49
and ARMA model, 49
data stationary, 49
initialization/fitting, 51
plotting, 50–52
predictions *vs.* actuals, 52, 53
RMSE score, 53
test predictions, 52
Autoregressive models (AR), 127
Autoregressive moving average
(ARMA) model
actuals, 47
ARIMA function, 43
bitcoin price data, 45
concept, 43
initialization/fitting, 47
libraries, importing, 44
loading data, 44
plotting, 45, 47
predictions *vs.* actuals, 47, 48
preprocessing, 45
RMSE score, 48
test predictions, 47
train-test split, 45, 46

© Akshay R Kulkarni, Adarsha Shivananda, Anoosh Kulkarni, V Adithya Krishnan 2023
A. R. Kulkarni et al., *Time Series Algorithms Recipes*,
https://doi.org/10.1007/978-1-4842-8978-5

B

Best-performing model
  evaluation, 119
  plot prediction, 121
  test set, 119
  validation set, 123, 124

C

Comma-separated (CSV) file, 4, 6

D

Deep learning methods, 127
  See also GRU model; LSTM

E

Exponential smoothing, 127

F

Facebook Prophet model
  added regressors, 84
    data, 85
    fit data, 86
    forecast data, 86
    initialization, 85
    label and encode, 85
    train-test split, 85
  adjusting trends, 79
    changepoint_prior_scale, 79
    plot the output, 80, 81
  change points, 73

hyperparameter, 76, 78
  magnitude, 76
  plot, 74
  print, 75
holidays
  custom dataframe, 83
  future dataframe, 83
  initialize dataframe, 83
univariate time series, 68
  dataframe for
    forecasting, 70, 71
  import libraries, 68
  initialization, 70
  plot forecast, 71
  plot forecast components, 72
  read data, 69
  training dataset, 69

G

get_dummies method, 109
Grid search hyperparameter
  ARIMA model
    evaluation, 54
    initialization/fitting, 58
    predictions *vs.* actuals, 59
    RMSE score, 60
  arima_model_evaluate
    function, 55, 56
  plotting, 59
  test predictions, 58
  tuning, 56–58
GRU model
  analyze data, 137–139

build, 147, 149
evaluation, 151–156
import libraries, 136
preprocess data,
    141, 142, 144
read data, 136
train-test split, 146

H, I, J, K

Holt-Winters (HW) model
    ExponentialSmoothing
        function, 65, 66
    initialization/fitting, 65
    plotting, 65
    predictions *vs.* actuals, 66
    RMSE score, 66
    test predictions, 65

L

LightGBM model
    build, 114
    evaluation, 114
    validation set, 115
LSTM
    data_prep function, 132
    fit model, 134
    import libraries, 128
    initialization, 132
    missing data, 129
    model predictions, 134
    model summary, 133
    normalization, 131

normalize data, 130
normalize_fn function, 130
plot the predictions, 135
plotting_actual_vs_pred
    function, 135
read data, 129
time series data, 130
train-test split, 131

M

Machine learning (ML) regression
        algorithms
    collect data, 105
    import libraries, 105
    preprocess data, 106, 108, 109
    select features, 109
    time series, 104
    validation split, 110
Mean absolute error (MAE), 118,
        151, 163
Mean absolute percentage error
        (MAPE), 118
Mean squared error (MSE),
        118, 151
Moving average (MA)
    definition, 34
    libraries, importing, 34
    plotting
        forecast *vs.* actual, 37, 38
        time series, 36, 37
    preprocessing, 35
    reading data, 35
    rolling mean, 37

Multiplicative model decomposition
   air passenger data, 25
   components, 25
   data processing, 25
   libraries, loading, 25
   plotting, 26
   quadratic/exponential, 25
   seasonal component, 27
   seasonality, 25
   time series, 26, 27
Multivariate time series data
   Beijing pollution dataset, 10
   definition, 9
   libraries, importing, 10
   loading dataset, 10
   loading/exploring, 9
   parsing function, 10
   plotting, 12, 13
   PM2.5 concentration, 12
   preprocessing, 10
   relationship, 9
Multivariate time series, VAR model
   build, 100
   evaluate the model, 101
   import libraries, 96
   preprocess data, 97, 98
   read data, 96
   split dataset, 99
   stationarity, 99

N, O, P, Q

NeuralProphet
   build model, 159–163
   import libraries, 158
   preprocess data, 159
   read data, 158

R

Random forest model
   build, 116
   evaluation, 116
   validation, 117
Recurrent neural network (RNN),
      128, 164
   fit model, 166
   initialization, 165
   model
         summary, 165
   plotting_actual_vs_pred
         function, 168
   predictions, 167
Root-mean-square error (RMSE),
      48, 53–55, 59–60, 62, 64, 66,
      101, 118

S

Seasonal autoregressive integrated
      moving average
      (SARIMA) model
   initializing/fitting, 60
   plotting, 61
   predictions *vs.* actuals, 61, 62
   RMSE score, 62
   SARIMAX function, 60
   test predictions, 61

Seasonality
    definition, 15
    libraries, importing, 15
    plotting
        date *vs.* temperature, 16
        monthly box
            plot, 16, 17
        yearly box plot, 17, 18
    reading data, 15
    temperature dataset, 15
    tractor sales data, 18
        datetime series, 19
        formatting data, 19
        libraries, importing, 19
        monthly box plot, 20, 21
        plotting, 19, 20
        reading data, 19
    visualization
        adding methods, 28
        box plot, 30, 31
        data processing, 28
        libraries, importing, 28
        line charts, 29, 30
        loading data, 28
        output, 29
        pivot table, 28
SelectKBest, 109
Simple exponential smoothing
        (SES) model
    definition, 63
    initialization/fitting, 63
    plotting, 63
    predictions *vs.* actuals, 64
    RMSE score, 64

SimpleExpSmoothing
    function, 63
    test predictions, 63

T

Time series
    analysis/forecasting, 1
    demand/sales, product, 1
    recipes, 1, 2
    uses, 1
Time series, Auto-ARIMA
    analyze data pattern, 90
    build, 92
    evaluate model, 95
    import libraries, 87
    output, 93
    preprocess data, 88
    read data, 87
    stationarity, 90
    summary, 94
    test data, 94, 95
    train and test, 91
Time series objects
    air passengers
        dataframe, 2
        libraries, importing, 2
        Pandas, 2
        parsing function, 2
        plot output, 3
        read_csv, 3
        reading data, 3
    India GDP data
        CSV file, 4

Time series objects (*cont.*)
    dataframe, 4
    libraries, importing, 4
    plot output, 5
    reading data, 4
    retrieved object, 5
    store/retrieve, pickle, 5
  saving, 6, 7
Trends
  definition, 13
  libraries, importing, 14
  loding dataset, 14
  parsing function, 14
  plotting, 14
  shampoo sales dataset, 13

U, V, W

Univariate statistical modeling,
  33, 34, 67

Univariate time series
  analysis, 9
Univariate time series
  data, 33
  definition, 7
  forecasting, 33
  libraries, importing, 7
  loading/exploring, 7
  plotting, 8, 9
  reading data, 8
  temperature dataset, 7

X, Y, Z

XGBoost model
  build, 112
  evaluate, 112
  validation set, 113

Printed in the United States
by Baker & Taylor Publisher Services